Human Cloning in the Media

This book examines the cultural production of cloning, its transformation from science fiction to science practice, and how this plays out in the global arena. The authors analyse the controversies surrounding both 'therapeutic' cloning for stem cell research and 'reproductive' cloning. Case studes are used to illustrate key issues, including:

- the 'story-telling' involved in scientific accounts of the future;
- the image of the scientist, scientific expertise and institutions;
- the governance of science;
- the representation of women's bodies as the subjects and objects of biotechnology;
- the ways in which 'the public' is perceived and represented in the human cloning debate.

Drawing together insights from science and technology studies and media and cultural studies, this book offers a timely contribution to debates about the public communication of science and the status of scientific truth. This book will be a valuable companion to students and researchers of media, science communication, cultural studies, sociology and science and technology.

Joan Haran is a Research Associate in the ESRC Centre for Economic and Social Aspects of Genomics (CESAGen), at Cardiff University.

Jenny Kitzinger is Professor of Media and Communication Research at Cardiff University.

Maureen McNeil is Professor of Women's Studies and Cultural Studies at Lancaster University.

Kate O'Riordan is Lecturer in Media and Film Studies at the University of Sussex.

Genetics and Society

Series Editors: Paul Atkinson, *Associate Director of CESAGen, Cardiff University*; Ruth Chadwick, *Director of CESAGen, Cardiff University*; Peter Glasner, *Professorial Research Fellow for CESAGen at Cardiff University*; and Brian Wynne, *Associate Director of CESAGen, Lancaster University*

The books in this series, all based on original research, explore the social, economic and ethical consequences of the new genetic sciences. The series is based in the ESRC's Centre for Economic and Social Aspects of Genomics, the largest UK investment in social-science research on the implications of these innovations. With a mix of research monographs, edited collections, textbooks and a major new handbook, the series will be a major contribution to the social analysis of new agricultural and biomedical technologies.

Series titles include:

Governing the Transatlantic Conflict over Agricultural Biotechnology (2006)
Contending Coalitions, Trade Liberalisation and Standard Setting
Joseph Murphy and Les Levidow

New Genetics, New Social Formations (2006)
Peter Glasner, Paul Atkinson and Helen Greenslade

New Genetics, New Identities (2006)
Paul Atkinson, Peter Glasner and Helen Greenslade

The GM Debate (2007)
Risk, Politics and Public Engagement
Tom Horlick-Jones, John Walls, Gene Rowe, Nick Pidgeon, Wouter Poortinga, Graham Murdock and Tim O'Riordan

Growth Cultures (2007)
Life Sciences & Economic Development
Philip Cooke

Human Cloning in the Media (2007)
From Science Fiction to Science Practice
Joan Haran, Jenny Kitzinger, Maureen McNeil and Kate O'Riordan

Genetically Modified Crops on Trial (2007)
Opening Up Alternative Futures of Euro-Agriculture
Les Levidow

Local Cells, Global Science (2008)
Embryonic Stem Cell Research in India
Aditya Bharadwaj and Peter Glasner

Handbook of Genetics and Society (2008)
Paul Atkinson, Peter Glasner and Margaret Lock

Human Cloning in the Media

From science fiction to science practice

Joan Haran, Jenny Kitzinger, Maureen McNeil and Kate O'Riordan

Routledge
Taylor & Francis Group
LONDON AND NEW YORK

First published 2008
by Routledge
2 Park Square, Milton Park, Abingdon, Oxfordshire OX14 4RN

Simultaneously published in the USA and Canada
by Routledge
711 Third Avenue, New York, NY 10017

First issued in paperback 2014

Routledge is an imprint of the Taylor & Francis Group, an informa business

Typeset in Garamond by
Keystroke, 28 High Street, Tettenhall, Wolverhampton

British Library Cataloguing in Publication Data
A catalogue record for this book is available from the British Library

Library of Congress Cataloging in Publication Data
Human cloning in the media / Joan Haran ... [et al.].
p. cm.
Includes index.
1. Human cloning in mass media. 2. Science in mass media. 3. Communication
in science. I. Haran, Joan, 1965–
P96.H84H86 2007
660.6'5—dc22
2007016947

ISBN13: 978–0–415–42236–9 (hbk)
ISBN13: 978–0–415–75966–3 (pbk)

Contents

Figures

Acknowledgements

The support of the Economic and Social Research Council (ESRC) is gratefully acknowledged. The work forms part of the programme of the ESRC Research Centre for Economic and Social Aspects of Genomics.

This book would not have been possible without the input and support of many people, to whom we offer very grateful thanks, and we attempt to account here for some of that wealth of contribution.

We are particularly indebted to all those who took part in our research; to the people who agreed to give up their time to be interviewed; to those who took part in focus groups, discussions, screenings and workshops; and to those who patiently filled in questionnaires. We thank all of these participants for their time, good humour, imagination and expertise.

In addition to the critical traditions with which we engage in the book, this project would not also have been possible without the exchange of ideas, criticism and support from colleagues both in the institutions where we work – Cardiff University, Lancaster University and the University of Sussex – and in the wider academic community. We would particularly like to thank Caroline Bassett, Choon Key Chekar, Fiona Coyle, Julie Doyle, Lena Eriksson, Emma Hughes, Callum MacKellar, Grace Reid, Celia Roberts and Jackie Stacey. We are also grateful to the staff at the Mass Observation Archive for their assistance.

Many of the ideas in this book were tried out at conferences and workshops, and responses to the papers that we gave were valuable in developing this work further. Organisations we have to thank in this respect include the Society for the Social Studies of Science (4S), the Media, Communication and Cultural Studies Association of the UK (MeCCSA); conferences and events in the Genomics Network, especially at the Genomics Forum; the annual CESAGen and Center for Society and Genomics conferences; WisCon; and the Society for Literature, Science and the Arts (SLSA).

Finally, we are indebted to the forbearance and support of our families and friends who have put up with us and helped us throughout this process.

1 Introduction

In the last years of the twentieth century and the early years of the twenty-first century, human cloning captured global media attention. Hailed as the source of potential cures for a wide range of human ills and feared as a violation of nature and an abuse of human beings – cloning has been the subject of news reporting in the UK, the USA and South Korea, as well as in many other parts of the world. It has also featured in Hollywood films, in television drama-documentaries and in notable best-selling novels. There is nothing surprising about this. After all, this decade has also seen celebrated, but also controversial, staged public media events pertaining to cloning: the 1997 announcement of the cloning of Dolly the sheep and the first declaration of the 'completion' of the Human Genome Project in 2000. Due in no small part to these announcements, this has also been a period in which expectations about cloning have grown exponentially and dramatically.

This book sets out to investigate the phenomena noted above: the flurry of cultural productions and texts which appeared in the wake of these two important markers in the history of genetics and biotechnology, the changing expectations regarding human cloning, and the state of this technoscientific field in the early twenty-first century. Our particular interest is in tracing and analysing some of the processes of mediation which have been a crucial, though often neglected or misconstrued, feature of the making of genomics in its recent incarnations. The particular focus of our concern has been human cloning.

By 'human cloning' we refer to the creation of a cloned human embryo – whether for stem-cell research, or with the aim of creating a human baby. Human cloning is often presented differently depending on the intended outcome – with the cloning for stem-cell research being designated as 'therapeutic cloning' and that oriented for reproduction being labelled 'reproductive cloning'. Indeed, 'therapeutic cloning' is sometimes simply referred to as 'stem-cell research', without any link with reproductive cloning being acknowledged in the terminology. At other times therapeutic cloning is represented as a sideline to the central project of embryonic stem-cell research. However, this distinction between 'therapeutic' and 'reproductive' human cloning is important and we investigate its significance in this book. There is, as we shall show, a long-established tension between cloning as re-generation and cloning as re-production. Nevertheless, therapeutic cloning builds on the technologies of reproduction, and 'reproductive' cloning may be framed as a therapy for infertility, following the model of *in vitro* fertilisation.

We begin our introduction by situating this book within a strong body of insightful analysis of recent formations of biological technoscience which includes important cultural and media analyses. Hence, we will identify some of these key influences and draw attention to our indebtedness to other scholars for valuable concepts which we have borrowed and adapted.

We continue with a short overview of the research project which generated this book and sketch the parameters of the analysis offered here, including the time period covered and the scope of our investigation. We briefly review our materials, our methods and our approach to data. We also include a preliminary signalling of two key issues. The first of these is the foregrounding of the mediated nature of technoscience as a challenge to much scholarship within recent technoscience studies. The second is our concern to highlight global flows but also to address national specificities in the analysis of technoscientific developments. Finally, this introduction offers a brief overview of the structure of the volume and an outline of each of the chapters.

The team, the field and conceptual tooling

This book has been produced by four feminist researchers who were brought together to constitute the flagship project on Media, Culture and Genomics, working from two different sites (Cardiff University and Lancaster University) within the ESRC Centre for Social and Economic Aspects of Genomics (CESAGen). We were virtually strangers when we assembled in this cross-site team, but we came together in the process of interrogating recent developments in genomics and we focused on cloning as a particularly controversial aspect of recent biotechnological development. This research was, from the outset, very much a collective process. Our most substantial gesture towards a division of labour was an early decision that the Cardiff University team (Jenny Kitzinger and Joan Haran) would focus on news and documentary media and that the Lancaster University team (Maureen McNeil and Kate O'Riordan) would devote its energy to fictional and more 'imaginary' forms, particularly film. As we discovered and indicate later in the book, this division was, in many respects, thrown open by the research that emerged and, indeed, by our developing conceptions of the field. As the arguments of this book suggest, to have maintained a rigid division between our analysis of fact and fiction would have required distorting a particularly important dimension of the picture of cloning discourses that we identify (see especially Chapter 6).

As four researchers with disparate academic and personal backgrounds, our coming together facilitated the sharing of a rich repertoire of conceptual resources from many different fields. Our disciplinary heritage includes anthropology and history, as well as cultural, gender, literary, media, science,

science-fiction and women's studies. Moreover, we were eager to identify appropriate concepts, insights and framings that would help us during our somewhat unconventional and unwieldy journey. Because our 'findings' include these findings – these very useful borrowings – we will briefly outline some of them here. They should also help to orientate our readers for the rest of the book.

In shifting our project away from an unwieldy tracking of the proliferation of genomic discourses in the media to a more specific focus on cloning, Evelyn Fox Keller's (Keller 1995, 2000) work on twentieth-century biology and the development of genetics was very influential. Keller's overview of these biosciences encouraged us to think that exploring cloning in the era after the mapping of the human genome would be a valuable and distinctive project. She has suggested that, in the early twentieth century, the emerging discipline of genetics – in part through the introduction of the distinction between genotype and phenotype – distinguished the problem of hereditary transmission from that of embryological development, marking a disciplinary split in biology between genetics and embryology (1995: 5). Developments in genetics and the mapping of the human genome at the end of the twentieth and beginning of the twenty-first centuries, combined with some successful initiatives in *somatic* cell nuclear transfer in mammals, have opened up the prospect of bringing together these two disciplines again, as part of the pursuit of therapeutic interventions. In a sense, Keller's work provides the conceptual framework that illuminates the consequent transformation in cloning imaginaries. Indeed, the genealogies of cloning that we discuss in Chapter 2 indicate that specific understandings of the historical relationship between genetics and embryology are being produced and called upon to underwrite key contemporary biotechnological projects.

Steven Shapin and Simon Schaffer's *Leviathan* (1985) has been a landmark contribution to the history of science and science studies. These historians recast conventional understandings of the foundations for modern science in their detailed analysis of the struggles between Robert Boyle and Thomas Hobbes in the forging of arrangements for obtaining knowledge of the natural world. Shapin and Schaffer demonstrate that the making of modern science involved a very particular configuration of social and political structures in seventeenth-century Britain.

It may not seem obvious that this historical study should capture the attention of researchers investigating twentieth- and twenty-first-century genomics and the media. But Shapin and Schaffer's emphasis on the performative nature of modern science and their concern to delineate the technologies that enabled the enactments of science became crucial to our understanding of late twentieth- and early twenty-first-century bioscience. We took from them the idea that the generation of scientific knowledge has, in diverse and specific ways, involved a carefully regulated system of 'witnessing'. Part of our assignment was to investigate the specific ways in which such witnessing occurred and was contested in and through recent developments in genomic science. Shapin and Schaffer's foregrounding of the literary and social technologies which enabled such witnessing also intrigued us. Our pursuit of the features of more contemporary scientific literary and social technologies became another thread in our research. This was particularly the case since recent genomic science has been characterised by notable disputes about how these technologies function.

While *Leviathan* has been a landmark text within science studies, it has not remained immune from criticism. In a respectful engagement with Shapin and Schaffer's work, Donna Haraway (1997) has drawn attention to the specific historical conditions which contributed to the forging of the figure of the scientific witness. She insists that the story of the making of modern science must also be told as the story of the making of gender and invokes Elizabeth Potter's historical research to substantiate this claim (1997: 26–32). Haraway's proposal of the figure of 'the modest witness' as the ideal model of the modern scientist encapsulates the problematic gender relations of modern science. Her interrogation of this figure brings the realist epistemological orientation of modern science and the claims to legitimacy of its apparently disembodied and disinterested observers under scrutiny. We have borrowed Haraway's concept of the modest witness and we have found it very useful in interpreting the performances of contemporary bioscientists and their castings in the mass media.

Thomas Gieryn is another science studies scholar whose research has influenced our analysis. Gieryn's (1999) study of boundary making in twentieth-century science offers a useful model for understanding the public making of science. We have found his attention to the iterations and reiterations involved in this boundary work and his precise tracing of the role of the mass media in some of his case studies useful. In particular, we have employed his concept of 'second-order virtuality', which refers to the role the media play in ensuring forms of 'witnessing' in the development of new technoscientific fields.

We have also benefited from a wealth of other research. We situate this study within a now well-established and diverse tradition of feminist studies of technoscience. Indeed, feminist technoscience studies is now such a flourishing and complex field that it is difficult to fully acknowledge the multitude of ways in which it has contributed to and shaped this project. So, we merely mention here some key areas of research that we have drawn upon.

While we have already mentioned Donna Haraway's influence through the figure of 'the modest witness' (1997), her work has provided much broader inspiration for our research. We have noted her careful trackings of the circuits of socio-cultural production in her investigations of a range of technosciences, including informatics, primatology and genomics itself. We draw specifically on Haraway's mobilisation of the concepts of tropes and figures as we undertake our own tracings of modest and immodest witnesses, women's bodies and cloning technoscience (Haraway, 1997).

Feminist research on new reproductive technoscience has been crucial to this research project and we draw directly on some of this research in the following chapters. However, we are particularly indebted to the recent work of Karen Throsby (2004) and Charis Thompson (2005) who, as part of their own original research (mainly focused on IVF), offer important conceptualisations and distinctive reviews of feminist debates on reproductive technologies.

We were also drawn towards recent feminist cultural studies research which has moved on to the terrain of technoscience. We have been mindful of Anne Balsamo's insight that: 'all technologies reproduce cultural arrangements. In this sense, all technologies can be considered *reproductive* technologies' (1999: 87). Balsamo usefully summarises a tradition of thinking through the visual in feminist approaches. This work is salient to our investigations of the intersections of genomic technologies (particularly cloning), visualising technologies and the media. Hence we follow Valerie Hartouni (1997), Rosalind Petchesky (1987), Carol Stabile (1994), Lisa Cartwright (1995), Catherine Waldby (2000), Judith Roof (1996), Donna Haraway (1997) and others who have examined the complex vectors through which vision and technology intersect. The insights of these researchers have been important since we are concerned with how cloning and genomic technologies are seen and imagined.

This brings us to a final specific sphere of feminist work: research on genetics, genomics and what Sarah Franklin has designated as 'the genetic imaginary' (Franklin 2000). We see this book as continuing a tradition of feminist investigation of the meanings and political significance of the new genetics of the late twentieth and early twenty-first centuries. This research has been particularly important in locating bio-technoscientific developments within a broad cultural frame. Evelyn Fox Keller's research is important in this regard. In addition, Jose van Dijck's (1998) *Imagenation* and Franklin et al.'s *Global Nature, Global Culture* (2000) exemplify such framing. Jackie Stacey's (2003, 2005, forthcoming) more recent studies of the filmic life of the gene have continued this tradition, combining a broad cultural approach with more detailed studies of Hollywood films, the cinema's operations as a cultural form and recent developments in genetics.

The preceding feminist research is part of a larger body of work we would designate as cultural studies of genetic and genomic technoscience. Dorothy Nelkin and Susan Lindee's (1995) investigation of the iconography of modern genetics and of the impact of this new bioscience on daily life in the Western world was an early marker in such research. Jon Turney's (1998) powerful documentation of the Frankenstein myth as a key repository of twentieth-century images of technology was a touchstone for our project. We have also benefited from an extensive body of scholarship on genetics and genomics in film.[1] Research on discourse, rhetoric and policy pertaining to new reproductive technologies and genetics (Nerlich et al. 2002; Nerlich and Hellsten 2004; Nerlich 2004) also influenced our analysis. We draw particularly on Michael Mulkay's (1997) study of the debates around the regulation of new reproductive technologies and embryo experimentation in the UK.

Methods and methodology

This project was influenced by media and cultural studies conceptualisations and research that posited a complex cultural circuit, including production, texts and audience or readers (Gray 2003; Johnson et al. 2004). Our shared view was that meanings were not generated at one point of that circuit – to then radiate out. Rather, aware of the various and diverse locations for the generation of meaning at all points in the cultural circuit, we were concerned that our research should acknowledge and work with this awareness. We devised a three-stage strategy that was designed to engage us in textual analysis, followed by investigation of the production and producers of some of the key materials we had analysed. In the final stage, we wanted to undertake audience research, possibly extending this into some broader investigation of publics.

We were confronted with a wealth of diverse texts: articles in newspapers, television news items, Hollywood films, television drama documentaries and so on. It was clear that keeping track of this material would, in itself, be a major strand of our research. We evolved a strategy which combined comprehensive time-limited samples with the tracking of key issues across a broader time frame and across a wider range of media.

We already had access to a comprehensive collection of national UK newspaper and television news coverage for 2000 (collected for an earlier Wellcome Trust-funded project). We complemented this by building up a similar archive of UK news coverage for the first six months of 2004 and the first six months of 2006). As the project progressed we also collected television news and press coverage materials around key events that fell outside this time period, including, for example, international 'breakthroughs' in human cloning or major consultation or legislative initiatives.

A second substantial documentation component of our study was the compiling of a filmography. The appendix in this volume is the result of this work (see Appendix II). Viewing and analysing this film archive and related texts constituted a substantial part of the textual analysis of this project. In addition to contemporary films, we also viewed earlier films which were important for their representations of cloning, including *Invasion of the Body Snatchers* (1956), *The Boys from Brazil* (1978) and *Parts: The Clonus Horror* (1978).

A third substantial investigation involved tracking particular issues across a wider range of media and cultural events. We thus collected relevant television documentaries, dramas and drama-documentaries, analysed radio coverage and examined websites. We also attended scientific conferences and public engagement events, theatre and dance performances and gallery exhibitions. Our studies extended to a wide variety of other types of texts ranging from a national postage stamp to policy reports, and from consultation documents to scientists' own accounts of their work. Attention to the multiple sites across which key products, tropes, figures and stories operate is an important part of our research journey. We are interested in how all these cultural texts operate together, and in dialogue with one another, to constitute human cloning.

Some readers may be surprised by the way we have analysed accounts of scientific developments written by a leading scientist alongside a reading of a drama documentary, film or newspaper report. We have done this quite deliberately, in order explicitly to challenge the division between 'science' and mediation. We do not treat any particular text as essentially factual rather than fictional, objective rather than biased or 'consultative' rather than 'prescriptive'. Our interest is in how truth claims are asserted and refuted through particular processes and genres. Such an approach brings together science and technology studies' (STS) conceptions of scientific truth/practice with media/cultural studies attention to media discourses, production strategies and contexts. Our approach to the construction of fact/fiction draws on the strengths of both STS and media/cultural studies and, we hope, brings distinctive insights to both fields.

Late twentieth- and early twenty-first-century developments in the biosciences lend themselves particularly well to this approach because they are so often identified as challenging ideas about what is fact and what is fiction. Breakthroughs in this area of science are variously characterised as 'tearing up the textbooks', 'rewriting the book of life' and 'turning science fiction into fact'. Indeed, as we shall argue, in attempting to shift human cloning from science fiction to science practice, the scientific establishment has emerged as a 'dream machine' comparable in many ways to Hollywood. Moreover, this technoscientific domain has also been populated by protagonists who continually cross and re-cross any fact/fiction divide. Scientists identify themselves, or are identified by others, as pioneers, dreamers, charlatans, mavericks or fallen heroes at different times and in different places. Celebrity campaigners associated with such research similarly occupy liminal or fluctuating positions. Christopher Reeves, for example, was an actor who played Superman, who then became a frontman for stem-cell research, and then appeared (as himself) in an advert which was digitally manipulated to realise his fantasy of walking again. Another celebrity transgressing the fact/fiction divide in the cloning field is Arnold Schwarzenegger who played a clone in *The 6th Day* (2000) and also exploited his other fictional personas to become a political figure as the governor of California, a US state that has developed pro-active stem-cell policies.

Our critical approach to the fact/fiction distinction and analysis of a wide range of texts is combined with attention to the diverse processes through which they are produced. In order to explore this further we interviewed some of the key figures involved in representing human cloning. Our interviewees include scientists such as Ian Wilmut (of Dolly the sheep fame), Alison Murdoch (from the Newcastle team involved in cloning for stem-cell research) and Panos Zavos (the notorious, US-based scientist who claimed to have implanted a cloned human embryo into a woman's womb). We also interviewed campaigners such as Josephine Quintavalle (representing the UK-based

'Comment on Reproductive Ethics' – CORE) and Judy Norsigian (from the US-based Our Bodies Ourselves Collective – also known as the Boston Women's Health Collective).

We were also interested in interviewing media professionals involved in producing cultural representations of human cloning. This was harder to achieve. Overall, we have simply been pragmatically strategic in this part of our research. As we anticipated, although we tried, we were not able to make contact with the makers of Hollywood films but we did interview television producers, theatre directors, journalists and artists. We also attended panel discussions and workshops in which producers of science programmes discussed their work. While we would not claim that our examination of this dimension of cloning culture has been thoroughly systematic, we have extensively investigated this part of the cultural circuit and can map some key contours of this terrain.

Given our interest in the public making of science and the influential recent work within both media and cultural studies on audiences and readership, we were particularly keen to incorporate some audience research into our project. There were a number of components of this. We used various opportunities to undertake focus groups or engage in public forums investigating how audiences viewed cloning in films or television programmes which appeared during the period of our research. Hence, two of us participated in the Edinburgh Film Festival on the Ethics of Human Cloning in November 2005. We helped the organisers of that event, including assisting with the drawing-up of a questionnaire to which we added some of our own questions about media and cloning. At the WisCon Science Fiction Convention in May 2006, we led focus group discussions after showing the first episode (*Resurrection*) of the BBC television drama series *11th Hour* and an audience discussion after a screening of the film *Teknolust*. There was also a set of ten focus groups conducted in Cardiff and London during which research participants discussed their views on stem-cell research and were invited to write their own news bulletins using photographs taken from television news coverage. We organised workshops for school students in Lancaster in 2005 and 2006 which involved their viewing of clips from recent films which represented cloning, followed by discussions, an exercise in compiling imagined headlines for cloning news, and, in one case, art work.

Another research initiative oriented to soliciting information about how the public related to genomics and some of its recent representations in the media was undertaken in collaboration with the Mass Observation Archive at Sussex University. In its current form the Archive has a register of respondents who are contacted periodically to write briefly in response to a set of questions pertaining to contemporary issues or activities. The Archive staff set up a directive on this topic on our behalf which invited participants to write about how they learned about genes, genetics and cloning.[2] We received 174 responses to our directive. (These are kept in the permanent archive at the University of Sussex, and are available to researchers.)

Thinking and researching locally and globally

In undertaking this project we were all very aware of the global nature of the contemporary mass media and of the global networks that were emerging in genomic technoscience. Hence, we knew that the project would necessarily involve investigating this global casting of this burgeoning technoscientific field. At the same time it is clear that this is a field where the *national* stakes are high. In many respects the Clinton–Blair media events which marked the 'completion' of the Human Genome Project in 2000 were indicative of a much wider pattern. Clinton and Blair were declaring the HGP a global achievement that would have implications for the entire world, but their appearance also signalled the high national stakes (in this case, for the USA and UK) at play. We knew from the outset that we would need to be mindful of both these dimensions of the cloning story.

Our national foci were determined by practicalities and by developments in the field. We set out to make genomics and the media in the UK our prime concern. This was determined not only by our location and by resource limitations, but also by the role the UK was assuming as a leader in the field of human genomics as it emerged in the late twentieth and early twenty-first centuries. This can be signalled by reference to obvious markers in the field: the cloning of Dolly in the Roslin Institute in 1996 and the media announcements about her birth in 1997; the contribution by John Sulston's team in the mapping of the human genome; and the Clinton–Blair press conferences of June 2000 mentioned above. Equally significant for this project was the groundswell of attention being given to human genomics and cloning, especially in the UK mass media during this period. We were also aware of institutional developments in the UK (for example, the elaboration of the responsibilities of the Human Embryo and Fertilisation Authority) which signalled that it was positioning itself to play an important role in this field and that investment in it was moving up the political agenda.

While the focus on the UK appeared an obvious choice, we thought it important to supplement this with research on other national contexts. This probably would have been required if only to give us some handle on what was actually happening in the UK. That is, if Britain was positioning itself as one of the front-runners in a race (and, indeed, this is often the term in which it has been described), we had to know something about the other runners. In addition, the collaboration/competition of the UK and USA in the HGP made some work on US developments seem crucial. However, there were more specific and practical reasons for building in research about other national developments into the project. Hollywood films loom large in any project on mass media and biotechnology in the contemporary world. This meant that we would necessarily be engaged in researching US culture. Moreover, our networks and contacts in the USA in science, media, cultural and gender studies all pulled us in this direction.

Thus, our initial sense of the project was of one which would mainly focus on developments in genomic technoscience and the media in the UK, with attention to related and contrasting developments in the USA. The dramatic developments which brought South Korea into the frame, particularly from 2004 onward, transformed our research. We originally identified the reported breakthrough in February 2004 as a significant news event and therefore collected all reporting related to Hwang Woo-Suk (for a brief explanation of Hwang's significance, see the last section of this chapter) in the national UK press and television news. In addition, we attended a conference in Edinburgh at which Hwang spoke and we interviewed him there. After the subsequent breakthrough announcement in 2005, further interviews were conducted in South Korea on our behalf by Choon-Key Chekar. She interviewed South Korean journalists and representatives from Korean religious and civic organisations who were challenging some of Hwang's work. These interviews included a representative from the Catholic Church in South Korea, from the feminist group, Womens Link, and from the People's Solidarity for Participatory Democracy group.

As the two breakthrough news events were followed by more extensive questioning, first of Hwang's ethics, and then of his science, we also worked closely with Choon Key Chekar on collecting and coding an extensive archive of Korean press coverage for the weeks of the breakthrough of February 2004 and May 2005, and each week around key moments in the subsequent scandal of November and December 2005. In addition Lexis–Nexis searches were conducted for the weeks of the four key news events for all the UK national press, and also for *The New York Times* and the *Washington Post*. Relevant articles in *Science* and *Nature* were also analysed. At the same time, our on-going interviews with key international stem-cell scientists and attendance at conferences allowed us to monitor reaction across the scientific community.

As the foregoing outline indicates this book is based on an intense study of developments around cloning and the media in the UK over the period between 1996/7 and 2006/7. We have also undertaken some analyses of related developments and patterns in the USA and South Korea. We would not pretend that this is a comparative study. This would not have been possible given the time, resources and scope of our project and team. However, we hope that we have been able to signal some important features of human cloning science, including key aspects of its mediation, in the USA and South Korea. We have been mindful of international dimensions and of global networks and institutions both in mass communications and in contemporary biotechnology. We have no illusions about having provided a comprehensive review of the global dimensions and structures of recent genomic technoscience, but we hope we have made some interesting suggestions about it.

Finding your way: the chapter guide

The body of this book is made up of six chapters, each addressing a different set of core issues. Chapter 2 poses the overarching question: 'What is cloning?' Posing this question at the start of the book is our way of signalling that there have been important shifts in the meanings of this term during the past decade. We start to answer this question by critically examining a set of genealogies which have been offered in key publications that have appeared since Dolly's birth in 1996 (see Appendix I for a related chronology). Our critical readings suggest that there is no single origin story of human cloning but also that there are no neutral accounts of this field. The juxtaposing of these stories also helps us place our own investigation on a larger and more complex canvas. It establishes crucial background for the second exploratory trajectory in this chapter: our review of some of the crucial changes in the cultural imaginary of cloning during the last decade. The final component of Chapter 2 is the brief overview of the discursive terrains of cloning in the specific national settings we researched: UK, USA and South Korea. This includes a sketch of important legislation pertaining to cloning and the identification of the key discursive actors in debates on cloning and related genomic technoscience. We also incorporate an account of developments around generating international legislation in this field.

In Chapter 3 we consider the temporal dimensions of recent cloning discourses. Here we trace the particular and peculiar fixation on cures as the promise of recent cloning technoscience. We trace the discursive temporal condensations that frame and encourage these expectations. More generally we trace the changing visions of cloning futures as these have emerged over the past decade.

The next two chapters address human agency and corporeality within the recent discourses of human cloning. In Chapter 4 we analyse the figure of the cloning scientist who has been so central to this emerging area of bioscience. We alight on familiar images of the maverick and madman not only because these are such well-established tropes in Western media, but also because these types have been so strongly conjured in recent developments in cloning science (for example, in the figuring of Panos Zavos who claimed to have attempted reproductive cloning). The story of Hwang and his team, who became the standard-bearers for South Korean biotechnology and for that nation more generally, forms the focus of our investigation of a 'fallen hero'. The final element in this chapter involves the gendering of the cloning cultural imaginary, as we provide a brief review of the problems of picturing women scientists in this field.

While analysing the representations of scientists seems an obvious component for a study of the making of a public science such as genomics, the inclusion of a chapter on women's bodies may not seem such an obvious choice. However, as we make clear in Chapter 5, women's bodies have been at the centre not only of the material process of somatic cell nuclear transfer, but also of the making of cloning science more generally during the last decade. Hence, we trace the very striking patterns of women's appearance and disappearance that have been a feature of this mediated field.

In recent years, scientists have made concerted efforts to take the media seriously. While there are a few celebrity scientists and designated 'good communicators', there is still much unease about interaction with the media. Science studies as an academic field has been even more timid and/or dismissive about addressing the media. While this whole book addresses these lacunae, Chapter 6 attempts to fill out the picture by paying much more attention to genres and media conventions. It does this while investigating the fact/fiction divide which is continually reiterated around human cloning. Here we investigate news values and their impact on the emerging field of cloning. We also undertake some case studies of what we call liminal media forms: a very unusual television drama-documentary on cloning (called *If . . . Cloning Could Cure Us*), the website of a popular film on cloning (*Godsend*) and the websites of a maverick cloner (Panos Zavos). This chapter challenges monolithic understandings of the media and, through some detailed analyses of cloning in specific media forms, interrogates the relationship between truth claims and genre.

We turn from genre to publics in Chapter 7. Here we try to specify and give substance to the understandings of how publics are figured and placed in the making of recent genomic science. There have been intense debates in the UK, the USA and South Korea about public engagement with science during the period of our research project. This chapter considers how publics have been conjured in the making of cloning technoscience during the last decade.

In our concluding chapter we offer some reflections on our analysis of human cloning and the media since 1996. Our commentary is presented as a set of responses to common clichés associated with human cloning that have circulated in the late twentieth and early twenty-first centuries. As part of this, we return to the powerful but unstable distinctions that have characterised this discursive territory: therapeutic and reproductive cloning; fact and fiction.

Having outlined some of the background and features of our research project and having provided a guide to its organisation and structure, we now invite our readers to follow our account of the making of human cloning in late twentieth- and early twenty-first-century media.

2 What is cloning?

> We should not see cloning as an isolated technology, single-mindedly directed at replication of livestock or of people. It is the third player in a trio of modern biotechnologies that have arisen since the early 1970s. Each of the three, taken alone, is striking; but taken together, they propel humanity into a new age – as significant, as time will tell, as our forebear's transition into the age of steam, or of radio, or of nuclear power (Wilmut et al. 2000a: 6).
>
> The point is that the three technologies together – genetic engineering, genomics and our method of cloning from cultured cells – are a very powerful combination (ibid.: 9).

Introduction

Ian Wilmut and his colleagues make dramatic claims for cloning as a technology. However, cloning captured popular attention long before news of the birth of Dolly the sheep in 1997 and before the announcement of the 'completion' of the Human Genome Project in 2000. Nevertheless, in the aftermath of these two announcements there has been a renewed and intensified interest in cloning. This chapter brings the contemporary concern with human cloning under scrutiny both by locating it in a set of genealogies and by analysing the distinctive features of its recent manifestations. (See Appendix I for a timeline that attempts to consolidate these genealogies to provide an overview of key events.) In this chapter, we draw on a range of accounts by scientists and social commentators on the development of the technology of cloning and provide a brief review of some of its key representations in Western popular culture. This sets the context for our analysis of the particular configurations of cloning discourses that have become dominant in the early twenty-first century.

It is tempting at this point to offer a dictionary definition of cloning as our starting point. For example, the *Chambers Dictionary* published the following definition for clone:

> *n* a group of two or more individuals with identical genetic makeup derived, by asexual reproduction, from a single common parent or ancestor,

> *orig* applied to plants, but later applied much more widely; any of such individuals; a person or thing closely similar to another, a copy or replica (*colloq.*). – *vt* to reproduce as a clone; to produce a clone or clones of ... [Gr. *klon* shoot]
>
> (1993: 324)

While this definition indicates some contemporary resonances of the term, it is our contention that, in the early twenty-first century, cloning has accrued a new set of meanings, associations, imagery and iconography that render this definition inadequate. This chapter provides an orientating introduction to the research on specific aspects of recent controversies about cloning that are explored in the rest of this book. It does this by offering both a longitudinal analysis of the diverse strands in the genealogy of cloning and a set of reflections about cloning in the contemporary cultural imaginary. Before moving on, it is worth noting that the practice of using Greek roots in the coinage of terminology can mislead the casual reader as to the antiquity of a term. The word 'clone' was not coined until the early twentieth century. In 1903, Herbert J. Webber of the US Department of Agriculture invented the term: 'to describe a colony or organisms derived asexually from a single progenitor' (Silver 2001; Webber 1903). Silver points out that Webber's coinage 'found quick acceptance among botanists and gained favour among biologists working with cells in culture' (2001).

In the discussion that follows we highlight some of the key boundary demarcations that have characterised more recent representations of cloning, focusing particularly on the distinction between human reproductive cloning and therapeutic cloning that was instantiated early in the post-Dolly discourse.[1] This account reviews a complex intersection of technoscientific developments, within genetic engineering, the Human Genome Project (HGP), and assisted reproductive technologies, all of which can be subsumed under the more general title of biotechnology/ies. But we also draw attention to other sites in which cloning is constructed and given meaning. Towards the end of this chapter we briefly outline the key elements of current legislation pertaining to the practices of cloning which have emerged in the countries which have been the focus of our research – the UK, the USA and South Korea. We draw attention to some of the implications of the distinctive national patterns in the global circulation of discourses about genomics, cloning and stem-cell research. We supplement this with a sketch of recent initiatives to provide international legislation and regulation in this field. Our review of specific national and international developments also includes the identification of the key discursive actors in recent public debates about cloning.

The genesis of human cloning

In the wake of the announcement of the birth of Dolly the sheep in 1997, journalists, bioethicists, academics from a wide range of disciplines and scientists, including those most directly involved with Dolly, have generated a wealth of interesting and interested accounts of cloning, focusing variously on the history, the future and the technological, social and ethical implications of this scientific 'breakthrough'. Lee Silver, a biologist at Princeton University, was famously quoted in Gina Kolata's breaking of the Dolly story in *The New York Times* saying: 'It's unbelievable. It basically means that there are no limits. It means all of science fiction is true. They said it could never be done and now here it is, done before the year 2000' (Kolata 1997). The rush to print popular accounts of the breakthrough and its implications was one indicator of the international stir that the announcement evoked. The titles of some of the most popular of these accounts proclaimed that Dolly's birth marked a new age for cloning, including: *Clone: the Road to Dolly* (1998) by *New York Times* science correspondent, Gina Kolata and *The Second Creation* (2000) by Ian Wilmut and Keith Campbell, the scientists credited with Dolly's birth. Indeed, the sea-change in scientific and public opinion that Dolly's birth is claimed to have occasioned is signalled in the title of the more recent popular science offering by Wilmut – *After Dolly* (2006). The latter volume, co-authored with Roger Highfield, the science editor of the *Daily Telegraph*, suggests one genesis point for the notion of human cloning as a plausible future prospect. Wilmut's reading of the history of cloning and of scientific achievement in this field is clearly not a neutral one and it centres on his achievements and those of his collaborator Keith Campbell at the Roslin Institute in Scotland. Nevertheless, the confrontation with human cloning as a potentially imminent technological prospect has become a main feature of the discourse on cloning in the last decade. This was indicated by moral philosophers Martha C. Nussbaum and Cass R. Sunstein when they offered the following evaluation of the significance of Dolly's appearance:

> the world reacted with intense emotion . . . the arrival of Dolly made it clear that human beings would soon have to face the possibility of human cloning – and it has been this idea, far more than the reality of animal cloning, that has caused public anxiety.
>
> (Nussbaum and Sunstein 1999: 11)

Genealogical tracking: 'a zoo of species'[2]

Although we would not advocate the same narrative trajectory, the descriptive chronology set out in *After Dolly* merits review – not least as the book was published simultaneously in the USA and the UK and addressed to a mass market readership.[3] It thus provides a key popular resource for understanding the history of cloning from the vantage point of the early twenty-first century, ten years after the announcement of Dolly's birth. Wilmut and Highfield celebrate somatic cell nuclear transfer, which we discuss below, the key technology for Dolly's creation, as the focus of contemporary scientific promise and they offer a genealogical account which identifies a narrow linear scientific ancestry for cloning.

The lineage thus constructed begins with Hans Driesch's work in the late nineteenth century on sea urchin embryos, which he successfully split and then allowed to develop into multiple (four) individuals. This work, Wilmut and Highfield note, demonstrated the principle on which current hopes for stem-cell research and therapy depend, that: 'a few divisions after fertilisation, embryonic cells retain the ability to turn into any type, from heart cells to egg or sperm cells, or even a whole individual (biologists say that they remain "totipotent")' (Wilmut and Highfield 2006: 52).

The next key contributors to the field that Wilmut and Highfield cite, for offering 'the first glimpse of the nuclear transfer process that we used to create Dolly' (p. 53) are Jacques Loeb and Hans Spemann. Jacques Loeb showed how to simulate fertilisation in sea urchin eggs. Hans Spemann, who they suggest is 'the true father of cloning', worked with salamander eggs. A 1914 account explains that Spemann manipulated salamander eggs to produce effectively enucleated cells which were then re-nucleated with the nucleus from one cell of a sixteen-cell embryo. Spemann was therefore working with undifferentiated cells, but, in 1938, he proposed a 'fantastical experiment' which would replace the nucleus of an egg with that of a differentiated cell, although he never carried it out (p. 55).[4] This step was realised by Robert Briggs and Thomas King (1956), who transferred the nucleus of a frog cell taken from an early embryo into another frog's egg.[5] Using this procedure, in 1951 'they successfully reconstructed frog embryos that grew into tadpoles' (p. 56). However, their experiments with embryos at different stages led them to conclude that it would be impossible to produce a clone from the nucleus of an adult cell (p. 57). This conclusion was overturned by the 1966 work of John (now Sir John) Gurdon who had some success in producing frogs through introducing the DNA of a cell nucleus from a juvenile animal (a tadpole rather than a frog) into enucleated eggs. Wilmut and Highfield quote Gurdon: 'This key experiment justified the view that the cloning of differentiated, and perhaps even adult, cells was at least theoretically possible' (p. 58). The credibility of his work was questioned, but a 1975 paper, according to Wilmut and Highfield, provided 'convincing evidence that Gurdon really was using fully differentiated donor cells' although he was never successful in cloning an adult frog from an adult frog cell (pp. 59–60). Wilmut and Highfield suggest that 'the story of mammalian [as opposed to amphibian] cloning can be traced back to one of Gurdon's pupils, Chris Graham' (p. 60) and his development of a 'gentle' nuclear transfer technique for mice. Further developments in the microsurgical equipment and culture techniques needed for nuclear transfer they credit to Derek Bromhall who worked with rabbit eggs (p. 62–3). They discuss the disputed, and largely discredited, work of Peter Hoppe and Karl Illmensee, who claimed in 1981 to have cloned mice. They suggest that Hoppe and Illmensee's research, however problematic, was important in the creation of Dolly, because: 'his [Illmensee's] lectures would inspire two key figures in her story, Keith Campbell and Steen Willadsen' (pp. 71–2). Further, they contend that, in attempting, unsuccessfully to replicate Hoppe and Illmensee's results, other experimenters, such as Davor Solter and James McGrath developed new protocols for nuclear transfer.[6] Steen Willadsen is the final scientist discussed in this account of the development of mammal cloning technology. In 1986 he announced in *Nature* that he had cloned sheep from early embryos and Wilmut and Highfield pronounce that: 'Willadsen's lambs were the first mammals of any kind to be cloned – beyond any doubt – by nuclear transfer' (p. 78). However, as they also explain, it was learning in a private conversation that Willadsen had successfully transferred a differentiated cell that inspired Ian Wilmut to begin the programme of research that led to the creation of Dolly. Wilmut then visited Willadsen to discuss his work. According to Wilmut:

> He was not only fun and hospitable, but also a gentleman (when scientists say that someone is a gentleman, or generous, they mean that he openly discusses his ideas and reveals those apparently insignificant details of experimental method that are crucial for success). ... He gave me plenty of useful advice. Later he showed me the technical details.
>
> (p. 82)

What made Dolly a breakthrough?

In February 1997, seven months after her birth, news of Dolly's existence leaked out, secrecy having been maintained while the paper accounting for her 'creation' was being refereed for *Nature*. Despite the fanfare that ensued, Dolly was not the first 'cloned' mammal. In *Clone*, Gina Kolata suggests that an 'atmosphere of mostly blind indifference' had greeted Wilmut and Campbell's earlier cloning success, despite the publication of their paper on the birth of Megan and Morag (the cloned forerunners of Dolly, generated at Roslin) in *Nature*. This was, she argues, because 'few molecular biologists paid much attention to research involving farm animals' (Kolata 1998: 180–1). Richard Holliman gives a rather different account, claiming that: 'science journalists would have been sensitised to the newsworthiness of cloning experiments, following the reporting of Morag and Megan' and that 'this may go some way to explaining why the first UK article to report on this experiment was published prior to the end of the press embargo' (Holliman 2004: 119).

What made Dolly's creation a 'breakthrough' was the fact that the nucleus of the embryo from which she developed was that of a fully differentiated cell – an adult, or a somatic cell. Previous to this there had been other mammals, including other Roslin lambs, which had been born through asexual reproduction. These had been the result of techniques involving the replacement of the nucleus of an egg or oocyte with that of an embryonic or foetal cell, rather than through fertilisation of an egg 'naturally' or through combining egg and sperm in a laboratory. Dolly's birth demonstrated that a cell which had already specialised for a particular role in an organism could be reprogrammed once transplanted into an egg, making it possible to clone an embryo from the cell of an animal which had already reached maturity and whose salient characteristics were therefore known. This was important for the work that was being undertaken at Roslin which involved genetic engineering of large mammals to produce protein drugs for humans. Beyond this, the inference drawn by the media and other commentators confronted by Dolly's birth was that it made the cloning of humans a genuine and probably imminent technological possibility. Nussbaum and Sunstein's commentary quoted above is, in this sense, indicative of the expectations for human cloning precipitated by Dolly's birth.

It is notable not only that Wilmut and Highfield make Dolly's generation and somatic cell nuclear cloning the telos of their account of the history of cloning, but also that they refer *only* to research on non-human embryos. As Wilmut and Highfield point out, the cloning of Dolly 'relied on a zoo of species' (Wilmut and Highfield 2006: 48). This reference to zoology and the detailed account they provide of what they call a history of 'human-assisted animal cloning' separates the genealogy of cloning from the history of the development of human reproductive technologies, particularly IVF. It also instantiates a distinction between the reproductive cloning Wilmut had conducted with animals and the therapeutic cloning that he now hopes to conduct with human embryos. As we shall argue below, the marking and reinforcement of this boundary (between reproductive and therapeutic cloning) has become a crucial feature of recent cloning discourse. More immediately, we turn to the innovations around human reproduction in the last quarter of the twentieth century to offer a fuller assessment of the significance of these developments in relation to Dolly's generation and the conceptualisation of cloning in the wake of her appearance.

Assisted reproduction, micromanipulation and the history of IVF

Despite Wilmut and Highfield's account which foregrounds experimentation with animal embryos, cell nuclear replacement technology with human cells could not have been developed without input from a quite different trajectory of research. CNR was facilitated not just by experiments on animal embryos, but also by the innovation of *in vitro fertilisation* (IVF) as a 'therapy' for human infertility. *The Second Creation* and *After Dolly*, as well as Gina Kolata's *Clone,* draw attention to the importance of micromanipulation technologies and expertise, crucially developed in the assisted-reproduction field, in making cell nuclear transfer in microscopic mammal oocytes a possibility. Earlier experiments with cell nuclear transfer (or 'transplantation' as it was sometimes called) had used amphibian eggs. According to Kolata:

> Their eggs are huge: a frog's egg is almost two millimetres in diameter, visible to the naked eye. A human egg, in contrast, is ten times smaller,

> microscopic in size, and its volume is much less than a tenth that of a frog's egg.
>
> (Kolata 1998: 37)

Of course, another key difference is that amphibian embryos develop outside the body, and securing their eggs (oocytes) does not require invasive 'harvesting' techniques.

Hence, the genealogy of cloning and the prospects for human cloning today are linked to the history of the development of human reproductive technologies and to the evolution of practices associated with IVF during the late twentieth century. From this perspective, the birth of Louise Brown in 1978 was as significant a genealogical marker in this field as was the birth of Dolly the sheep in 1996. The attention given to the development of micromanipulation techniques by Wilmut and Highfield acknowledges some aspects of this legacy, but Kolata provides a fuller account of the history of IVF as a crucial strand in the history of cloning.

We would elaborate on the genealogical accounts considered here by arguing that the development and wide-scale implementation of IVF has not only been crucial in the development of the *technological* capacities for cloning but that it has also facilitated the emergence of a *cultural* imaginary for cloning and, as we will show below, a *legislative* framework (at least in the UK) for cloning applications. In her recent detailed study of women's experience of IVF failure in the UK, Karen Throsby makes the more general argument that IVF has come to constitute the core technology for 'the newer, more controversial reproductive and genetic technologies . . . in relation to which IVF is increasingly being constructed as normalised and unproblematic' (Throsby 2004: 187). She views the legitimisation that IVF has increasingly provided for subsequent developments in reproductive and genetic technologies as contributing to a process she calls 'technological creep' (ibid.: 189). Throsby's argument highlights the way in which IVF has been a key touchstone which has offered the social legitimacy and reassurance that have facilitated developments in this field. It is this, together with it providing the context for the development of micromanipulation techniques, which makes the development of IVF a crucial contributory branch in the genealogy of cloning. We shall explore the legal framework it also provided later in this chapter.

Thus far we have reviewed some of the significant genealogical accounts of cloning which have appeared in the aftermath of Dolly's creation in 1996 and used these to demonstrate the emergence of the perception that human cloning had become an imminent technological possibility. Our reading of the historical accounts of 'the Road to Dolly' is critical and we have stressed that Wilmut and Highfield tend to concentrate on one genealogical thread – animal embryological experimentation.[7] In contrast, we have emphasised the importance of developments within human reproductive technologies and the legitimising role of IVF in the recent history of human genetic technoscience. Nevertheless, thus far we have concentrated our attention on technological genealogies. Since we contend that it is not possible to understand contemporary conceptualisations of human cloning without considering other stories about its history, we now turn to some of these.

Other stories about cloning's genealogy

The genealogy provided by Wilmut and Highfield which we have sketched above is similar in many ways to that which had been provided by Gina Kolata in *Clone: the Road to Dolly* (1998). This text was one of the first and most celebrated accounts of the background to Dolly's birth and Kolata was *The New York Times* science correspondent who broke the Dolly story in the USA. Although there are some commonalities in their genealogical sketching, *Clone* and *After Dolly* differ in significant ways. *After Dolly* focuses on the centrality of the work done at Roslin in the realisation of the potential of two of the most notable biotechnological projects of the twentieth century, genetic engineering and the Human Genome Project. Kolata, in contrast, represents the work at Roslin as a pragmatic sideline to high-status microbiology and she identifies its positioning outside of the scientific mainstream as precisely what facilitated the achievement of Dolly. Beyond this distinction, Kolata's review of the history of cloning is much more encompassing, as she outlines the ebb and flow of expectations regarding cloning on a much wider canvas of scientific and general social controversy. Understanding the changing pattern of scientific and popular cultural expectations regarding cloning is an important part of Kolata's version of the story of cloning.

Gina Kolata characterises the work leading up to the cloning of Dolly (as outlined above through reference to Wilmut and Highfield's account) as coming 'from a tradition of utter romanticism' and the desire 'to understand the abiding biological mysteries of development and the psychological mysteries of identity' (Kolata 1998: 36). In other words, she envisages cloning as a matter of scientific imagination, as well as technical capabilities. She contends that scientists' preoccupation with cloning as a concrete technological prospect first emerged in an era when scientists were less subject to what she calls 'the organised ethics movement' (p. 70) that emerged towards the end of the twentieth century. She maintains that, for this reason, at that time, their vision of human cloning was more utopian than was the case in the late 1990s (p. 60). Kolata refers to futuristic speculation by the British biologist, J. B. S. Haldane, in a speech published in 1963 and by the American microbiologist and Nobel laureate, Joshua Lederberg, in articles published in 1966 and 1967 that heralded human cloning as potentially beneficial to humankind (pp. 61–2). Scientific expectations in relation to cloning were at this point, Kolata claims, highly speculative and she comments that: 'Today, most scientists would scoff at the leaps of faith necessary to contend that those experiments with frogs could possibly mean that human cloning was a real possibility' (p. 62–3). Nevertheless, she indicates that Haldane's and Lederberg's speculations emerged from a social context of generalised optimism about the ability of science to improve the conditions of human life in the Western world.

Kolata subsequently turns her attention to another phase in scientific interest in cloning. Less than a decade later, in 1971, James Watson, of double helix fame, suggested that the rapid progress in 'test-tube conception' by Patrick Steptoe and Robert Edwards meant that concerns about human cloning needed to be tackled urgently if governance of embryo research and its reproductive and therapeutic implications was to be undertaken in accord with the wishes of the public in the various societies where it might be conducted, noting that: 'Until recently, however, this foreboding [that Gurdon's work with frogs could be done with human cells] has seemed more like a science fiction scenario than a real problem that the human race has to live with' (Watson 2000: 83).[8] Watson argued against leaving important decisions pertaining to the implications of 'in vitro human embryo experimentation' – such as surrogate motherhood and cloning – either only to the government in Britain (where the relevant experiments were being undertaken at that time) or solely in the hands of the scientific and medical communities. This was because, he argued, not every scientific advance would 'automatically make our lives more "meaningful"' (ibid.: 89). Watson's article is particularly interesting because of its explicit identification of IVF research as the likely source of the techniques which would render the cloning capacities developed with amphibians applicable to humans. Moreover, Watson's reference to a 'science fiction scenario' alludes to another cultural source for ideas and imagery of cloning which we take up below.

Kolata maintains that: 'Watson's sounding of the alarms fell on almost deaf ears. Certainly, he did not mobilise the citizenry to start a cloning debate' (Kolata 1998: 71). However, she also points to the establishment of the Hastings Center,[9] which was founded to foster and undertake research on biomedical ethics in 1969. She notes that the Center's founders, Willard Gaylin and Daniel Callahan, publicised their new ethics institute with an article about cloning which appeared in *The New York Times Magazine* in March 1972.[10] However, both Kolata and Gaylin, whom she interviewed for *Clone,* reflected that Gaylin's theorising about clones:

> was met with stony coldness by many scientists. Even those who did not dismiss human cloning out of hand as being scientifically impossible decided that it would be so inane to clone humans that we need not fear it would come to pass.
>
> (Kolata 1998: 75)

In fact, Kolata suggests that, in the USA at least, 'a new wave of theological and ethical speculation about cloning' did not arise until the late 1970s and that this was ushered in through the 1978 publication of David Rorvik's *In His Image: The Cloning of a Man*. In this book, Rorvik claimed that he had helped a multimillionaire clone himself. He also drew narrative links between the animal cloning experiments of Bromhall and the efforts of human IVF pioneers in the USA through mention of the physician Landrum Shettles, who had previously claimed to have created the world's first *in vitro* pregnancy with a laboratory-fertilised human egg (Henig 2004: 35). Kolata characterises Rorvik as 'a respectable science writer' and his publishers, J. B. Lippincott as reputable. Peter Poon suggests, however, that: 'the publishers' preface, while facially [*sic*] neutral was provocative if not disingenuous: "The account that follows is an astonishing one. The author assures us it is true. We do not know."' (Poon 2000: 168). Three months after its publication, the Oxford scientist, Derek Bromhall, launched a libel suit against both writer and publisher. Bromhall's experiments with rabbits were cited in Rorvik's book and Bromhall did not wish to be seen as underwriting the credibility and accuracy of the book.[11] Paradoxically, Poon suggests that this was what made *In His Image*:

> [both] so popular and popularly excoriated . . . Rorvik relied on the science of cloning as it had been developing (and with which he had become well acquainted) to lend an air of convincing verisimilitude to his book . . . In contrast to the literary and cinematic productions that preceded his own work, Rorvik placed his 'real life' scenario squarely within the confines of the possible.
>
> (Poon 2000: 167–8)[12]

In 1982, Bromhall was granted a legal award for damages and he received an apology from the publisher, conceding that it believed the book to be untrue (Kolata 1998: 118).

Kolata's genealogy makes it clear that the history of cloning encompasses far more than the history of key technological innovations in the field of animal embryology and reproduction. In addition to amplifying the importance of the history of IVF in the trajectory to Dolly's generation, she sketches bioscientists' changing expectations around cloning. She demonstrates that, from early in the twentieth century, the idea of scientifically engineered human cloning has engendered fears and expectations but also, at some points, loathing and disinterest within the world of science and in Western culture more generally. Moreover, she shows that during the course of the twentieth century, bio-experimentation in cloning repeatedly became controversial: contested and/or mired by fraudulent claims. Her exposition also underscores that, from the 1970s, the prospect of human cloning became closely linked to the evolution of new institutions and mechanisms for governance of the biosciences as Watson's clarion call in 1971 and the founding of the Hastings Institute for Bioethics in 1969 indicate. We appreciate Kolata's expanded genealogy of cloning, which portrays its development as a matter not only of technical achievement but also a matter of scientific and popular cultural imagination. However, we are wary of her sceptical construal of what she calls 'the organised ethics movement'. We also disagree with her dismissal of critical perspectives on technoscience which she tends to present merely as obstructions that had to be overcome on 'the road to Dolly' and on the more long-term progressive march of the biotechnologies.

Cloning: other sites and other genealogies

In rather different ways, Kolata and Wilmut and his co-authors acknowledge that cloning has accrued meanings and resonant imagery outside of the professional scientific arena. All of these authors deploy literary and filmic imagery in their reviews of the history of cloning and in mounting their arguments about its recent prospects. There is, in fact, a long cultural history of the representation of cloning in literature and film (particularly in the genre of science fiction) which informs and enriches contemporary Western debates about and imagery of cloning. We see these as important sites in which cloning has been made and re-made. Hence, this book will offer a sustained analysis of recent filmic treatments of cloning. However, at this point, we simply draw attention to this long and influential cultural trail within popular literature and film – highlighting some of its key texts.

Although Mary Shelley's *Frankenstein* (1818) did not revolve around cloning, it offered an early and, what is now considered to be a classic, science-fiction image of the 'artificial' creation of (quasi) human life. Hence, it is often identified as part of the Western cultural imaginary associated with fears of cloning. Jon Turney (1995) argued persuasively that the Frankenstein narrative has provided the dominant imaginative framing for the popular representation of developments in genetic science during the twentieth century. It is probably no coincidence that Wilmut and Highfield refer to *Frankenstein* early in *After Dolly* (2006). Nevertheless, Aldous Huxley's *Brave New World* (1932) is the first science-fiction classic which furnishes explicit images of cloning and, as such, it has had considerable popular resonance ever since its initial publication. Nonetheless, neither 'clone' nor 'cloning' make an appearance as terms in *Brave New World*:

> Since the term 'clone' was reserved for botanical usage early in the [twentieth] century, one author [Huxley] invented the fictional phrase 'Bokanovsky's Process' to describe the manipulation of a human egg such that it 'will bud, will proliferate, will divide. From eight to ninety-six buds, and every bud will grow into a perfectly formed embryo, and every embryo into a full-sized adult. Making ninety-six human beings grow where only one grew before. Progress.
>
> (Huxley 1939: 3, cited in Poon 2000: 162)

These two literary texts have been long-term reference points and sources of imagery that continue to be drawn upon in recent discussions and representations of cloning. Although exactly how such fictions influence public imagination is often misrepresented (see Chapter 7).

In addition to these classic texts, there was a flurry of publication of science-fiction short stories and novels about cloning by leading authors in this field early in the 1970s. So, even if 'the citizenry' had not been mobilised by James Watson's call to debate, the topic of cloning was garnering attention within the genre of science fiction during this period. In 1975, Naomi Mitchison, sister of the scientist J. B. S. Haldane, published *Solution Three*, which indicated that she was rather less sanguine about the utopian social potential of cloning technology than her sibling. In addition, several cloning novels appeared in 1976. Amongst these were Pamela Sargent's *Cloned Lives* and Kate Wilhelm's *Where Late the Sweet Birds Sang*. Sargent's text explores the perspectives of each of the clone offspring of a single adult male at different points in their life course. But more significantly it also locates the conduct of the experiment in a well-drawn sociological account of the intersections of science, society and the media. In this way, it exemplifies what Peter Poon has called 'science unfiction': 'the *process* by which scientific developments both shape and are shaped by the imaginative projections of the nonscientific community' (Poon 2000: 160). Less rigorously grounded in near-future extrapolation than Sargent's account which opens on New Year's Eve 1999, Wilhelm's book is set in a post-apocalyptic world and explores the prospect that cloning would lead to a loss of human individuality and creativity. She also uses the cloning theme to explore anxieties about totalitarian societies and the threats of environmental degradation and nuclear devastation. However, it was Ira Levin's novel, *The Boys from Brazil*, published in 1976, which would become the best-known cloning text of this period – mainly through its transformation into the script for a Hollywood film.

In fact, the 1978 Hollywood production of *The Boys from Brazil* marked the beginning of a sequence of films which generated and circulated a repertoire of imagery which has been crucial in the framing of cloning in the late twentieth and early twenty-first centuries. In later chapters we trace and analyse this imagery as it emerged in films of the last decade. However, at this point, it is important to register that, from its release in 1978 to the present, *The Boys from Brazil* has been a crucial and persistent point of reference in conjurings of human cloning. There were other films on the theme of cloning made during the 1970s, including *The Clones* in 1974 and *Parts: The Clonus Horror* in 1979, but they lacked the international distribution of *The Boys from Brazil* and hence failed to gain either much recognition at the time, or the latter film's lasting notoriety. Nevertheless, this cluster of films does indicate the ongoing cultural exploration of cloning within Western popular film and together they constitute part of the reservoir of cultural imagery which has been drawn on in the post-Dolly period.

We cannot offer an analysis of science fiction as a genre here, nor of all the science fictional treatments of cloning prior to the birth of Dolly, although there are a number of insightful scholarly overviews available, as well as many scholars of technoscience who draw on the resources of science fiction.[13] Nevertheless, we take from these sources and from our own readings in this field an understanding that this is a genre which is profoundly engaged with modern technoscience and with the investigation of its role and significance in modern Western societies. While this encompasses a variety of dispositions towards technoscience, the representation of human cloning within twentieth-century science fiction has predominantly expressed ambivalence about such a practice, with fears and anxieties generally attached to its social and ethical implications rather than to the safety or efficacy of the technology per se. This textual ambivalence, however, means that interested parties can make a range of different, partial (in both senses of the word) claims about the messages conveyed about cloning by science fiction. As Michael Mulkay has usefully pointed out in his discussion of the embryo debate in the UK Parliament in the 1980s:

> when people speculate about the development of new, science-based technologies, they cannot rely entirely on what they take to be the established facts. In thinking and arguing about the shape of things to come, they have no alternative but to create some kind of story that goes beyond these facts.
>
> (Mulkay 1997: 117)

Nonetheless, as he goes on to point out, supporters of such new technologies seek to undermine the testimony of their critics by 'removing their arguments from the factual realm to an unreliable non-factual domain of supposition, fantasy and mere belief' (ibid.: 126). They routinely do this by claiming that their critics have been misled by the pernicious and irrational influence of science fiction.

The referencing of the literary and filmic heritage of cloning imagery in popular accounts of cloning since Dolly's appearance thus constitutes a revealing but complex cultural practice. For it is clear that commentators such as Wilmut and Highfield or Kolata do see the cloning of Dolly as profoundly disrupting the boundaries between the imaginary and, it is often suggested, the illegitimate scenarios of science fiction and the future, legitimate prospects encompassed by the technoscientific imaginary. We have already drawn attention to references to the prospective shifting of this boundary in Watson's essay on 'Clonal Man'. Kolata registers a more material shift: 'Until Dolly entered the world, cloning was the stuff of science fiction' (Kolata 1998: 3). Indeed, as we will demonstrate in more detail in later chapters, invoking the contrast between science fiction and contemporary technoscientific scenarios registers the shift in cloning discourse in the recent period: that, since Dolly's birth, many see human cloning as an imminent technological possibility. Hence, recent disavowals of the science-fiction tradition by some scientists and others committed to realising technoscientific human cloning are also a way of denying the dangers and dispelling anxieties associated with it by insisting on its technoscientific promise. However, the powerful legacy of earlier (nineteenth- and twentieth-century) Western science-fiction literature and film still casts its shadow across more contemporary discourses of cloning. Those who are optimistic about the potential of human cloning hence often adopt an ambivalent position in relation to science fiction. On the one hand, like Wilmut and Highfield, they draw extensively on the rich repertoire of cloning imagery generated in this important Western tradition. On the other hand, they often disavow it, insisting that 'science-fiction' cloning is the 'other' which must be distinguished from and not sully contemporary technoscientific prospects for cloning. Kolata's chapter titles are illuminating in this regard. The chapter entitled 'Imagining Clones' focuses on speculation by scientists, while the chapter dealing with Rorvik's hoax text, *In His Image*, with side discussions of novels by Mitchison and Levin is called 'Sullying Science'. It is likely that the history of fraudulent claims and discredited texts (particularly Rorvik's) associated with scientific experimentation in cloning earlier in the twentieth century as outlined above (and which has continued into the recent era, see Chapters 4 and 5 in particular), which is often regarded, effectively, as another form of 'science fiction', also contributes to this distancing impulse. As we will discuss later, the policing of the border between fact and fiction (whether the latter is acknowledged as science fiction or labelled science fraud) has become an important feature of the representations of this field of biotechnology since Dolly's birth.

Our exposition thus far has entailed borrowings from, and commentaries on, some of the most popular accounts of the history of cloning that appeared in the aftermath of the cloning of Dolly the sheep. We have used these to highlight four distinct but inter-connected genealogical strands that inform recent discourses of cloning:

- the history of technoscientific experimentation in animal embryology;
- the development of human reproductive technologies, particularly IVF;
- the changes within the scientific and popular cultural imaginary (expectations) around cloning;
- the history of the representation of cloning in literature and film (particularly within the genre of science fiction).

It is our contention that these are not just different ways of telling the story of cloning: they also indicate that the meanings of cloning do not emanate from one point of origin.

Our multiplicity of genealogies has also highlighted some important features of recent discourses of cloning, including:

- the expectation that human cloning is now a potential technoscientific prospect;[14]
- the strong imperative amongst those advocating technoscientific cloning to police the boundaries between fact and fiction (even though they may draw on literary and filmic imagery).

In the following sections of this chapter we re-visit points that emerged in our critical reading of recent histories of cloning that require further amplification to highlight other features of the contemporary discourses of cloning. We will consider, in turn:

- the focus on somatic cell nuclear transfer (SCNT);
- the proliferation of SCNT imagery;
- the emergence and reinforcement of the distinction between therapeutic and reproductive cloning.

Somatic cell nuclear transfer: cloning in the twenty-first century

In modern biotechnology, the term cloning may refer to many different kinds of processes and these can be carried out at the molecular or cellular level, as well as at the level of the organism as a whole. In fact, molecular cloning has been crucial in the evolution of genetic engineering and in the Human Genome Project. These are the technological fields that, together with mammalian cloning, Wilmut and his colleagues predict will realise the biotechnological convergence that could propel humanity into a new age. However, during the past decade, cloning has been increasingly identified with somatic cell nuclear transfer (SCNT) exemplified by the replacement (albeit partial) of images of multiple identical humans with images of eggs being enucleated and nuclei being transferred. This obscures the embryo splitting or non-somatic cell nuclear transfer versions of whole organism cloning that predated Dolly's birth,[15] as well as the molecular cloning that was crucial to the realisation of the Human Genome Project. *Somatic cell nuclear transfer* (SCNT) or *somatic cell nuclear replacement* (SCNR) is the technique used to produce the embryo that resulted in the birth of Dolly the sheep (the first mammal cloned from an adult cell) and cited as the telos of Wilmut and Highfield's history of cloning.

The following description of SCNT is our synthetic version of the potted textual descriptions of this process that can be found in many recent press reports and popular science texts:

> Somatic cell nuclear transfer or replacement requires the removal of the 'nucleus' containing the genetic material – the nuclear DNA – of a donor egg and its replacement with the nucleus of a differentiated cell. Through chemical and/or electrical stimulation, the egg is activated, and an embryo is thus created outside of sexual reproduction.

Figure 2.1 demonstrates a particularly well-drawn illustration of the way the process has been visually represented in newspaper coverage of the prospects of therapeutic cloning. In the illustration, human reproductive cloning is acknowledged as another potential application of the technology. In fact, the illustration accompanies a particularly reflective account of the contested

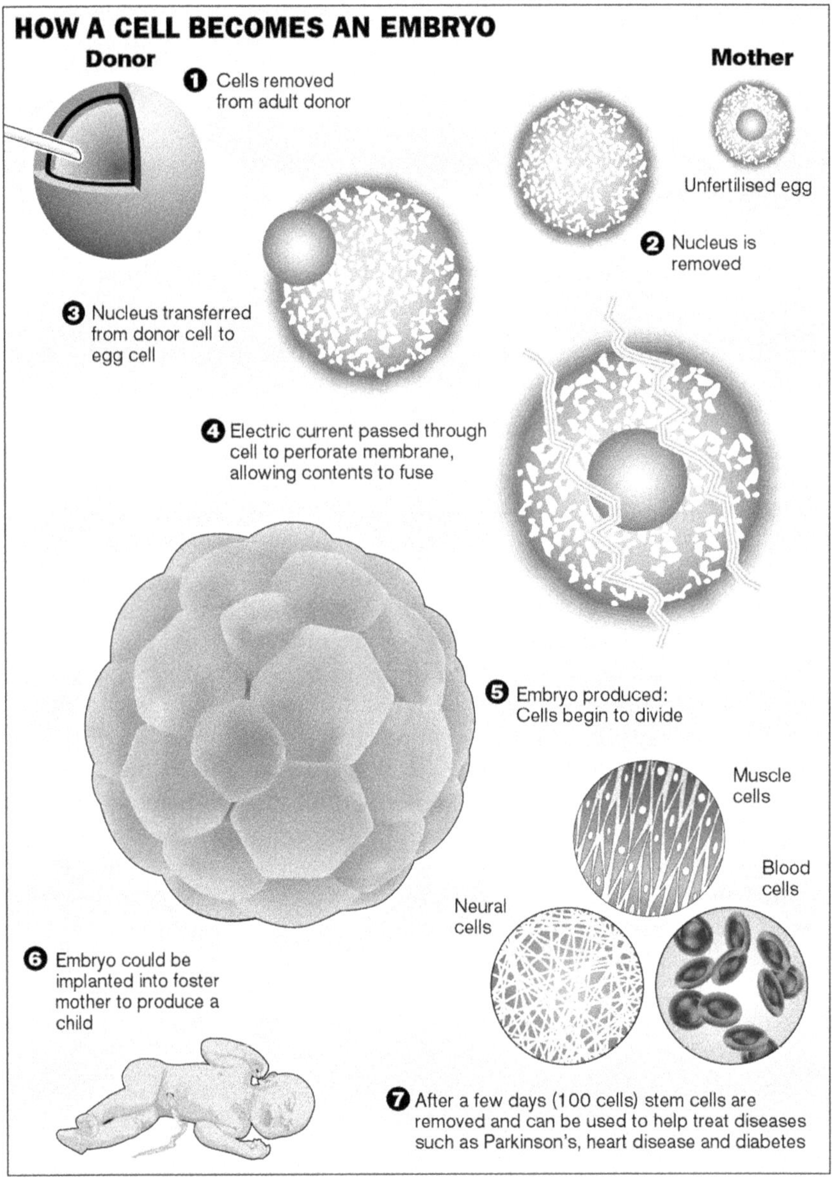

Figure 2.1 This illustration, 'How a cell becomes an embryo', appeared on page 5 of the *Independent* on 13 February 2004.

boundaries of reproductive and therapeutic cloning, which opens with the hook: 'The scientist behind the world's first cloned human embryo has admitted that his research techniques could help maverick doctors to produce cloned babies' (Connor and Arthur 2004).

As Keith Campbell and Ian Wilmut make clear in *The Second Creation* (Wilmut et al. 2000a), the type of simplified account of the somatic cell nuclear transfer process that we have provided, and that constitutes the level of explanation provided in most news media accounts, does not adequately represent either the biological transformation or the human intervention this process entails. The detailed account they provide of the development of cell nuclear transfer technology is intended to provide their reader with the competence to understand their research letter to *Nature* about the creation of Dolly. This letter succinctly describes the complicated cell culturing processes required to produce a cloned embryo. This is not to suggest that there is any deliberate misrepresentation in such explanations, but, rather that, in packaging the information in a way that appears to convey its essentials, important technical details have been elided. These elisions do not simply compromise technical comprehension; they have significant implications for discussions of the social and ethical implications of this process, because the simplification can lead to misunderstandings – particularly of the degree of difficulty and risk involved in this process. For example, this simple account does not register the likelihood that manipulating oocytes will lead to developmental problems in any embryos thereby produced. Wilmut has expressly stated that he is against human reproductive cloning, so the risk of deformed foetuses and miscarriages resulting from transfer of a (damaged) cloned human embryo to the womb of (for example) an infertile women is one that he would wish to see prohibited by law. But the issue of whether risks of damage to the embryo might also bear on therapeutic cloning applications is one that receives little public discussion.

In their paper describing the experiment which led to Dolly, Wilmut and Campbell coined the term 'reconstructed embryo' to distinguish the blastocysts they created in the laboratory from those resulting from sexual reproduction (Wilmut et al. 2000b: 313). This distinction is not one that has generally been employed in media discourses about cloning. Typically, the blastocysts produced using SCNT are simply referred to as embryos, a term which can obfuscate the distinction between the SCNT products and those embryos created in infertility clinics through IVF for the purposes of assisted reproduction. Such IVF embryos are also used for stem-cell research (with the consent of the IVF patients). Sarah Parry suggests that, in the UK at least, this confusion serves the interests of those in favour of conducting cell nuclear transfer research with human eggs, as the principle of permitting experimentation on embryos was established in the 1990 Human Fertilisation and Embryology Act. Had such advocates drawn attention to the difference between 'reconstructed embryos' and IVF embryos, they might not have been able to take advantage of the precedent set in the 1990 legislation. We discuss this and other aspects of this legislation in more detail below.

The increasing identification of cloning with SCNT or SCNR and the discursive focus on this process since Dolly appeared has been complemented and reinforced by the proliferation of visual imagery associated with it. Visual representations of the embryo, and particularly of the embryo in the process of enucleation, courtesy of photographs provided by the Roslin Institute, have become ubiquitous in news media coverage of stem-cell research and therapeutic cloning, making this the iconographic template of cloning in the early twenty-first century.[16] The particular illustration we have chosen to accompany our text (Figure 2.2) is available copyright-free for the use of non-profit organisations from the website of the Roslin Institute but similar images have also been circulated by Hwang's University of Seoul laboratory, and Advanced Cell Technology, amongst other organisations involved with SCNT research. Moreover, as we will illustrate in more detail in Chapter 5, both high-tech photography and line illustrations have been used extensively to 'explain the science' of therapeutic cloning.

Although it is the replication and wide currency of this image which is crucial here, it is interesting to note that visual representations of cell nuclear transfer or replacement predate its recent iconographic flourish. Given the

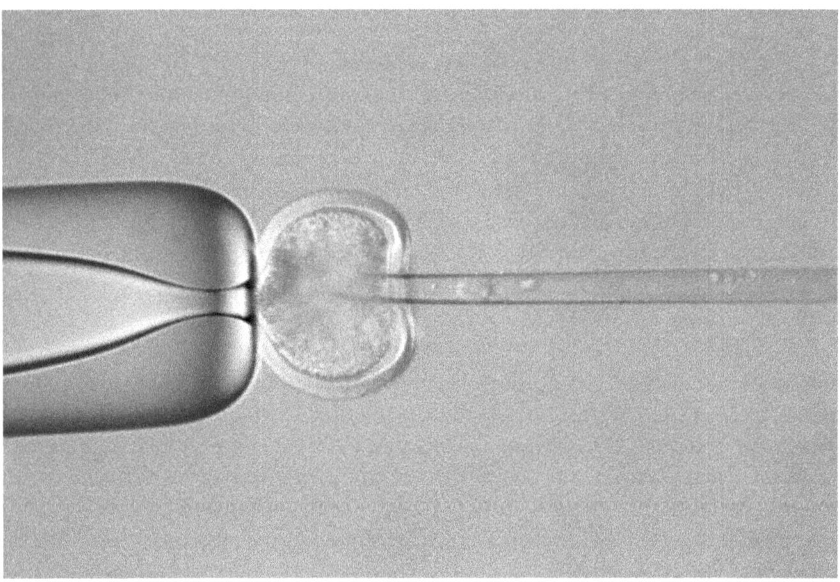

Figure 2.2 This image, provided by Roslin Institute, is captioned as follows on the Institute's website (www.roslin.ac.uk/imagelibrary/popups/107.php): 'Removing the maternal nucleus during nuclear transfer'. The web-mounted images can be used free of charge by education, public sector or non-profit making groups. This image, and others which are almost identical, has become iconic of the somatic cell nuclear transfer process.

recent rhetorical efforts to move cloning out of the science-fiction register discussed above, it is somewhat ironic that an early version of this image was featured in *The Boys from Brazil* (1978), the influential science-fiction film discussed above. This segment of the film demonstrates the micromanipulation of oocytes in footage that is visually virtually indistinguishable from the widely circulated contemporary stock images that are used as the televisual equivalent of Figure 2.2. According to film studies researcher David Kirby, this was because of the employment of Derek Bromhall of Oxford University (mentioned previously for his work on rabbit embryos and as a key figure in the Rorvik book controversy). Bromhall was the film's only scientific consultant and this, according to Kirby, influenced the presentation of 'Bromhall's conception of mammalian cloning as the only feasible model' (Kirby 2003: 55). Indeed, the imagery of SCNT/SCNR has also been employed in more recent offerings from Hollywood, such as *Godsend* (2004), a film which will be discussed in later chapters.

Reproductive cloning bad, therapeutic cloning good

As we have indicated, the boundary demarcation in human cloning between therapeutic and reproductive applications is a relatively novel discursive strategy. This strategy has emerged in the course of discussions about the implications of the birth of the first mammal cloned from a fully differentiated cell obtained from an adult donor. The widespread employment of the terms 'therapeutic cloning' and 'reproductive cloning' seems to imply and instantiate these as distinct technoscientific procedures. Nevertheless, both forms of cloning involve SCNT or SCNR as described in the preceding section. The distinction between them revolves around what happens after this process, since reproductive cloning would involve the placement of the resulting (reconstructed) embryo into a women's womb for gestation. Since the term 'therapeutic cloning' is increasingly used to refer to SCNT or SCNR, 'reproductive cloning' would be, in effect, 'therapeutic cloning' plus this placement. 'Maverick cloners' such as Panos Zavos challenge this distinction by suggesting that human reproductive cloning is, or would be, a therapeutic intervention for infertile couples.

Some claims have been made about successful reproductive cloning having already taken place, which we will discuss later in this book. However, as yet there is no credible evidence to support such claims and most have been roundly dismissed as hoaxes. Some limited success in therapeutic cloning has been reported – that is, the production of embryos using SCNT for the purposes of stem-cell research. However, again, these claims to success have been disrupted through allegations of fraud or premature publication. We will review the legal framework for these arguably distinct procedures in a later section of this chapter.

In an interview conducted at Roslin by Arlene Judith Klotzko (Klotzko 2003)[17] during the spring and summer of 1997, the chief scientists responsible for Dolly's creation, Ian Wilmut and Keith Campbell, both stated their moral opposition to reproductive cloning, but discussed the potential uses of therapeutic cloning for investigating and treating genetic diseases. Similar distinctions, between the potential benefits offered by the process of cell nuclear replacement and concerns over human reproductive cloning were registered by the House of Commons Select Committee on Science and Technology early in 1997 and these were then elaborated in a joint consultation document issued by the UK Human Genetics Advisory Commission (HGAC) and the UK Human Fertilisation and Embryology Authority (HFEA) in January 1998: *Cloning Issues in Reproduction, Science and Medicine.*

As Sarah Parry notes: '[W]hile human embryonic stem-cell research has been carried out since the mid-1990s, it was not until after the birth of Dolly the sheep that the [UK] Government formerly [*sic*] addressed policy issues surrounding related developments' (Parry 2003: 177). Referring to the HFEA/HGAC consultation process, Parry points out that:

> from this very early stage in the public debates, there emerged signs of the rhetorical severing of therapeutic cloning from reproductive cloning. Whilst it was recommended that reproductive cloning should remain illegal, and required primary legislation to ban it explicitly, therapeutic cloning was deemed to hold the promise of medical benefits.
>
> (ibid.: 179)

Parry's analysis of the stem-cell debates in the UK House of Commons and House of Lords demonstrates that these debates displayed what she calls 'discursive regularities' by which she means that they drew heavily on rhetorical resources developed by pro-research advocates in earlier debates on embryo research leading to the Human Fertilisation and Embryology Act (1990). This Act established the UK regulatory framework for the practice of IVF and for the creation, use, storage and disposal of any resulting embryos. Under this Act, research on embryos older than fourteen days was prohibited. Research was allowed in the period up to this limit but only through licence from the Human Fertilisation and Embryology Authority (HFEA). Licences were to be granted for research only in specified fields, mainly to do with fertility, reproduction and congenital disease.

Michael Mulkay suggests that there were 'three critical changes in [UK] parliamentarians' conception of embryo research' (Mulkay 1997: 132) which evolved over the course of this debate that lasted 'throughout the greater part of the 1980s' (p. 1). These three critical changes constituted successful boundary demarcation exercises. The first of these was the identification of the pre-fourteen-day-old human embryo as suitable experimental biological material, rather than as a (potential) human being. The second was 'an increasing emphasis on the possibility of controlling genetic disease with altering the genetic make-up of human individuals' (p. 133). In effect, this distinguished such genetic intervention from eugenics, by focusing rhetorically on the relief of suffering, when what was actually being offered, in the absence of any available genetic therapies, was the termination of particular pregnancies. The third was what Mulkay calls 'the success of the pro-research lobby [in] its replacement of its opponents' rhetoric of fear concerning the long-term impact of embryo research with an alternative rhetoric of hope' (p. 132–3).

So, although the birth of Dolly may have led to news reports framed within 'a rhetoric of fear' with regard to the possibilities of human *reproductive* cloning (see Nussbaum and Sunstein quotation on p. 15),[18] scientists and policy makers, in the UK in particular, already had a successful template of the future promise of genetic interventions on which to build. This allowed them to hold open the possibility of human cloning research by distinguishing between cloning embryos and cloning whole organisms. In 1998 the report that emerged from the UK HFEA and the Human Genetics Advisory Commission, following their public consultation on human cloning, recommended that consideration should be given to two further areas of embryo research: the development of treatments for mitochondrial disease and for diseased or damaged organs or tissues. The UK government then convened a group to review the field under the chairmanship of the then Chief Medical Officer, Professor Liam Donaldson. *The Donaldson Report* was published in August 2000 and recommended that research on embryos aim at increasing understanding about human diseases and disorders and developing treatments should be allowed. It also recommended that this research should be permitted on embryos created either by IVF or by cell nuclear replacement. These provisions were incorporated into legislation passed in December 2000. Kitzinger et al. point out that, during 2000, the British media continued to present the stem-cell debate as a dispute between two competing perspectives:

> On one side were those who felt embryonic stem-cell research was an abuse of embryos which set dangerous precedents (e.g. for reproductive cloning). On the other side were those who argued that the benefits outweighed any such ethical dilemmas or risk (if indeed such risks were seen to exist at all).
>
> (Kitzinger et al. 2007: 208)

Mulkay's reading of the previous debate on embryonic research makes it clear that a precedent had been set in the regulation of IVF that could be drawn on in this more recent debate on cloning. This relates to our argument about IVF providing the legislative framework for recent developments around cloning. However, Kitzinger and her colleagues show the rhetorical work that was required to maintain the distinction between points of view framed as 'science versus religion', as well as the flexibility of such discourses in responding to changing understandings of the terrain of debate. As technologies and social formations change over time, the discourses alter subtly in order to maintain their boundaries. Parry also points out that 'while the pro-research lobby had been slow in mobilising itself during the 1990 debates, in the cloning debates scientists and related institutions were more actively engaged in a pro-research campaign from the outset' (Parry 2003: 181). In the earlier debate 'the rhetoric of hope' was given purchase through the story of the 'baby without blemish' (Mulkay 1997: 69–82).[19] In the more recent debates on cloning, 'the rhetoric of hope' has been mobilised through the invitation to identify with patients who would be potential beneficiaries of new forms of research (Kitzinger and Williams 2005: 736). Drawing on the analyses of Mulkay, Parry and Kitzinger et al. together helps to illuminate the way that the discursive reiteration of 'the rhetoric of hope' in this debate about therapeutic cloning has been both recursive and performative. This demonstrates that, even though the embryo had been constituted as 'matter' for experimentation through the earlier IVF (1990) debate and legislation, it has constantly to be remade and performed as such.

Our detailed excursion into the debates about the regulation of cloning in the UK has underscored the discursive performativity of the distinction between therapeutic and reproductive cloning. As we indicated previously, therapeutic and reproductive cloning involve the same procedure, until the point of transport of the cloned embryo into the intended mother's womb. The labelling of SCNT as *therapeutic* cloning is also promissory: it presupposes its successful use in medical procedures. Nevertheless, as Wilmut and Campbell's declarations indicate, there has been considerable investment in this discursive distinction. Despite this, the technological overlap between therapeutic and reproductive cloning and, as we shall explain later, the difficulties in sourcing eggs except through *reproductive* procedures such as IVF, make this distinction unstable and require its continual reiteration.

More recently the boundaries between therapeutic and reproductive cloning have been assertively patrolled by pro-embryonic research advocates in the UK media. As we will go on to discuss in Chapters 4 and 6, the discursive work of maintaining the distinction between therapeutic and reproductive cloning has become linked to efforts to underscore other distinctions: between reputable and disreputable scientists, and between fact and fiction in cloning technoscience.

Naturalisation of reproductive cloning: embryo splitting and twinning

The post-Dolly period has seen some notable explicit attempts to naturalise or normalise technological cloning in publications emerging from the world of bioethics. Technological twinning and embryo splitting have been naturalised through arguments which invoke common-sense understandings of reproduction and the natural generation of twins and through the assemblage of repertoires of visual images of twins or identical copies. From our perspective in this chapter, this constitutes yet another way of telling the story of cloning in a way which seeks to counter arguments that frame cloning as artificial, dangerously novel and unnatural. In fact, prior to their discussion of the contribution to Dolly's cloning of a 'zoo of species', Wilmut and Highfield suggest that 'cloned humans have been around since the dawn of humanity' in the form of twins (Wilmut and Highfield 2006: 49).

One of the most prominent versions of such a naturalising representation of cloning is the British bioethicist John Harris's recent discussion of embryo splitting which draws an explicit mimetic analogy between the occurrence of identical twins 'in nature' and the deliberate splitting of *in vitro* embryos to create 'matching siblings' (Harris 2004: 3). He describes the splitting of pre-implantation (human) embryos in the laboratory as if this were a procedure that is currently readily available. Hall and Stillman reported the first instance of this artificial twinning in 1993, evoking some anxiety, but the technique's safety and efficacy for use in human-assisted reproduction is not yet established. Nonetheless, Harris cites the lack of take-up of this technology as evidence that there is no rush to take up new reproductive technologies associated with cloning. This is a somewhat disingenuous claim, since the procedure is prohibited in the UK at the present time. Moreover, there are serious questions about its safety.[20]

The medical ethicist Arlene Judith Klotzko normalises cloning in a somewhat similar way, although she underscores her points with visual imagery and humour. Her popular text on the science and ethics of cloning, *A Clone of Your Own,* which appeared in 2004, assembles a range of images of twins. Thirty-five illustrations are dominated by representations of twins (or larger multiples of siblings) including a Magritte painting, photographs of a television studio full of identical (monozygotic) twins and of the Dionne quintuplets. These images remind the reader that twins have generally been naturally produced and that they are a familiar and accepted part of Western culture.

Sarah Sexton of The Corner House[21] argues that 'proponents of human cloning often claim that the technique is nothing new because cloning occurs naturally in the form of identical twins, who result from an embryo dividing in two of its own accord during its early stages' (Sexton 1999). She counters this argument by pointing out that '[T]wins, however, have two genetic parents, not one, and are not genetically identical to either, obtaining half their DNA from each'. Her counter-argument demonstrates one of the difficulties of arguing by analogy as it also could be undermined using the same logic – since the nucleus in a cloned (SCNT) embryo also has two genetic parents, originally. However, there are distinctions that can be drawn between naturally occurring twins and those resulting from the application of SCNT cloning. These include the following: that the genetic contributions of the 'parents' are already one generation old at the time the embryo is activated; that the nucleus of the embryo is activated through chemical or electric stimulus rather than development being stimulated by the conjunction of the genetic contributions from sperm and egg, and that the mechanism by which this activation (or reprogramming) of the nucleus takes place is not fully understood. As Wilmut points out, with regard to Dolly:

> [She] is not an absolute, 100 percent replica of the old ewe who provided her first cell (who we might call her clone mother). She is not as similar to her clone mother as two identical twins would be. She is merely a genomic, or DNA clone.
>
> (Wilmut et al. 2000a: 45–6)

In fact, even this caveat does not capture the complexity of such 'replication', as there is genetic material outside the nucleus of each cell, and in the case of a clone produced through somatic cell nuclear transfer, this may or may not be provided by the same donor as the cell nucleus depending on whether or not she provides the oocyte (egg) that receives the new nucleus. In the case of Dolly, the 'egg donor' and the 'nucleus donor' were two distinct individuals.

However, the failure to register significant distinctions between embryo splitting and twinning on the one hand and cell nuclear transfer on the other facilitates discursive frameworks which naturalise/normalise the latter. In the process, any difference between spontaneous embryo splitting *in vivo* and technological splitting *in vitro* is also obscured. This is evident in the type of arguments employed by Harris and Klotzko, among others. Twinning and embryo splitting can be naturalised through appeals to common-sense understandings of reproduction and nature; they can also tap into an extensive Western cultural repertoire of visual imagery of the twin or copy.

Klotzko's extensive collection of images to support her popularisation of the bioethics of cloning begs the question of whether human reproductive cloning captures attention because it can be associated with a trope – twinning – which has an obvious and extensive visual register. This may be particularly attractive when so many of the recent developments in genetics and genomics are not easily rendered into a visual form.[22]

Of course, there is no guarantee that these texts will necessarily achieve these authors' intended goal. Because of the long history of twins being understood as disturbing and unsettling – associated with notions of the doppelganger and the uncanny – readers may not find these books reassuring. Nevertheless, both Harris's and Klotzko's texts are designed to naturalise and normalise cloning by associating it with forms of twinning which are already familiar in Western nature and culture. They address the benign image of the clone – the twinned human individual – and they seem eager to support the development of technoscientific cloning and assuage long-term fears of cloning posing a threat to human individuality.

Our discussion of these ventures in naturalising and normalising contemporary technoscientific cloning completes our review of recent discourses of cloning. We now turn to a different kind of mapping of the recent discursive domain of cloning – outlining the legislative context and introducing the key actors.

Overview of national and international discursive terrains: legislation and key actors

There have been substantial financial and imaginative investments in biotechnologies in each of the nation states we have studied. (This is elaborated in the next chapter, which examines visions of cloning futures.) The governance of cloning and of embryo research more generally and the concomitant legislation enacted to realise this has been under continuous review and these have been matters of heated contestation in different local and global contexts during the past ten years. Hence, we offer here a brief review of the legal status of cloning in each of the national settings that pertained at the time of going to press.

In the UK, human reproductive cloning was designated as illegal under the Human Reproductive Cloning Act of 2001 which includes a provision for up to ten years' imprisonment for infringement of this prohibition. As we have indicated above, therapeutic cloning is permissible, but only under licence from the Human Fertilisation and Embryology Authority. Therapeutic cloning and other forms of embryonic stem-cell research are generally publicly funded in the UK.

In many respects, South Korea has realised a pattern which is similar to that of the UK. Hence, in South Korea, reproductive cloning is also banned, while therapeutic cloning is permitted and such research has been underwritten through the commitment to substantial public funding: 'In 2004, the government allocated $5 billion (USD) to support research and has designated this work as a national priority. The Ministry of Science and Technology coordinates research, both private and public' (www.isscr.org/public/regions/country.cfm?CountryID=68, accessed 23 February, 2007). The Bioethics and Biosafety Act implemented in January 2005, some argue specifically for the purpose of facilitating the work of Hwang Woo-Suk, prohibits reproductive cloning and the creation of IVF embryos for non-reproductive purposes, but allows the use of supernumerary IVF embryos for research purposes, as well as the production of nuclear transfer embryos for research. In an article published in *The New York Times* following Hwang's branding as a fraud, Chin Kyo Hun, 'a professor emeritus at Seoul National University, who took part in drafting the law' was quoted as saying: 'The bioethics law had little to do with safeguarding bioethics but everything to do with giving Hwang a legal support' (Sang-Hun 2006).

In the USA the situation is more complex. At the national level, there has been no explicit legal prohibition of either reproductive or therapeutic cloning, although some states have enacted legislation banning reproductive cloning. Likewise, at the national level, there has been an embargo against the use of national funds for this type of research. Nevertheless, some states have allocated resources from their own funds to pursue therapeutic cloning and/or stem-cell research. The most high-profile instance of this to date has involved the State of California which passed Proposition 71 in November 2004. Through voting in favour of Proposition 71, Californians approved spending three billion dollars, raised through bond issues over ten years, for stem-cell research. Legal challenges issued subsequent to this Proposition have meant that none of the funds raised had been released as of December 2006. This impasse led the California Institute of Regenerative Medicine to seek funding through loans and grants from philanthropic institutions. According to its own website:

> The *California Institute for Regenerative Medicine* ('The Institute' or 'CIRM') is a state agency that was established through the passage of Proposition 71, the California Stem Cell Research and Cures Initiative. The statewide ballot measure, which provided $3 billion in funding for stem-cell research at California universities and research institutions, was approved by Californian voters on November 2, 2004, and called for the establishment of a new state agency to make grants and provide loans for stem-cell research, research facilities and other vital research opportunities/
>
> (http://www.cirm.ca.gov/faq/, accessed 23 February 2007)

Proposition 71 had been put on the ballot and campaigned for by the stem-cell advocacy organisation, Californians for Cures, which mobilised funds and endorsements from Hollywood celebrities, including Michael J. Fox, Dustin Hoffman and Brad Pitt and from Hollywood and other executives. Amongst these were the director and producer Jerry Zucker and his wife, Janet, who contributed over $50,000 and a great deal of time in support of the legislation (Usdin 2004). High-profile celebrity patient advocates have been significant players in the political process in the USA, particularly in providing endorsements and financial support for pro-research candidates.

At the global level, reaching agreement on whether and how to regulate cloning has proven virtually impossible because of fundamental disagreements over the principle of research on human embryos. On 6 November 2003, the Legal Committee of the United Nations General Assembly voted to suspend until late 2005 any decisions on an international treaty with regard to human cloning, initially prompted by a 2001 French–German proposal to ban human reproductive cloning (Center for Genetics and Society 2003; Harvey 2005). France and Germany had restricted their proposal to ban reproductive cloning, for strategic reasons, despite having prohibited research on therapeutic cloning in their own states. Other UN members, particularly the United States and Spain argued for a more wide-ranging formulation that would outlaw all forms of cloning with human embryos and therefore opposed the treaty as formulated. Such an expansion of the terms of the treaty was opposed by countries such as the UK, China, Singapore and Sweden with more permissive approaches to embryonic research and Belgium argued for revising the treaty language to prohibit reproductive cloning but to support research or therapeutic cloning. Despite the vote to postpone a decision on the treaty, wrangling continued at the UN resulting in the approval of a non-binding ban on all human cloning by eighty-four votes to thirty-four, with thirty-seven abstentions. Amongst the nations voting against the declaration were the United Kingdom and South Korea, with the USA being one of the most vocal supporters (Fifty-Ninth General Assemby of the United Nations 2005).

The coverage of evolving legal frameworks constitutes one strand in recent media discourses in the UK. Many news stories about human cloning, whether reproductive or therapeutic, evoke the Dolly template, typically associated with Ian Wilmut. However other key figures also garner attention, including Keith Campbell, Wilmut's collaborator in the sheep cloning experiments, and Harry Griffin, Deputy Director of the Roslin Institute in 1997 (and, at the time of writing, Roslin Director). More recently, news stories about an industrial tribunal by a disaffected former employee have brought the names of other Roslin scientists into the news frame (Harrell 2006). In 2004, Alison Murdoch and Miodrag Stokjovic, recipients of the first HFEA License to conduct therapeutic cloning using human eggs, became key figures in UK media coverage. Stokjovic has since left the Newcastle Centre for Life which holds the licence, moving first to Spain and subsequently to Serbia, but he continues to be mentioned frequently and quoted in the UK news media, and his movements signal the global importance of cloning research.

In news stories about the award of the Newcastle Centre for Life licence, as well as in later stories about successful cloning, other key actors include Suzi Leather, the Chair of the HFEA from March 2002 to August 2006 and Professor Stephen Minger, a US-born scientist working on embryonic stem-cell research in the UK. Minger's own work does not (at least, up to this point) involve therapeutic cloning with human eggs,[23] but he is one of a number of UK-based scientists who travelled to South Korea to witness the research on therapeutic cloning being conducted there. Suzi Leather was succeeded at the HFEA by Lord Richard Harries of Pentregarth, who took the position of interim chair of the authority until his replacement by Shirley Harrison in January 2007. Shirley Harrison is due to remain Interim chair of both the HFEA and the HTA (Human Tissue Authority) until their scheduled replacement (subject to legislation) by the Regulatory Authority for Tissues and Embryos (RATE) in the UK in 2008.

In South Korea, until recently, the key players in cloning were members of the team headed by Hwang at Seoul National University. In 2004, with the publication of what was greeted as a groundbreaking paper in *Science,* Hwang and his colleagues announced that they had cloned thirty human embryos and harvested stem cells from one of them (Hwang et al. 2004). This was followed, in May 2005, by a further announcement in *Science* that they had established eleven stem-cell lines derived from the skin cells of individual patients (Hwang et al. 2005). Although, as is common with scientific papers, both articles had multiple authors, Hwang as team leader was lionised in South Korea and courted by the international media, and scientists around the globe, until late 2005. Questions about Hwang's ethics emerged in the global media at this point and this eventually resulted in a full-blown scientific scandal. In early 2006 Seoul National University published a report that stated that the claims about research on cloned human embryos by Hwang and his team were fraudulent (Normile et al. 2006). Hwang has subsequently been subjected to criminal prosecution in South Korea on charges of fraud and embezzlement. The verdict on these charges was still outstanding as of March 2007. The ramifications of this case continue to unfold and its high profile means that Hwang remains a key discursive actor even though his scientific credibility has been catastrophically undermined.

In the USA, our third key national site, the bipartisan nature of national politics and a complex but powerful debate about science versus religion means that key discursive actors in the arena of embryonic stem-cell research and related areas such as therapeutic cloning are less likely to be high-profile scientists than is the case in the UK or South Korea. Although attention is given to some scientists such as Gerald Schatten (Hwang's US collaborator), the public face of the human cloning/stem-cell debates draws mostly on figures from the overlapping worlds of entertainment and politics. Celebrity advocates of such research have included Christopher Reeve, Michael J. Fox and Arnold Schwarzenegger. As Governor of California, Schwarzenegger was a powerful figure in the campaign to pass Proposition 71. Other supporters of human embryonic stem-cell and therapeutic cloning research have included Nancy Reagan, the widow of the former US president who had suffered from Alzheimer's, and John Kerry, the Democratic presidential candidate who unsuccessfully opposed George Bush Junior in the latter's re-election bid in 2005.

In the UK, the USA and South Korea, government leaders have been key discursive actors in the debates about cloning and stem-cell research. South Korean President Roh Moo-hyun was a strong supporter of the therapeutic cloning and stem-cell research, visiting Hwang's laboratory whilst he was still in favour internationally. US President Bill Clinton's instigation of the enquiry into cloning in the immediate aftermath of Dolly's birth and his joint statement with Tony Blair regarding the completion of the Human Genome Project were perhaps his most prominent interventions in this field. George W. Bush has identified himself strongly as an opponent of stem-cell research since taking office. Although Tony Blair has become a strong ally of this American president (e.g. over the Iraq war), he has taken a rather different stand on matters pertaining to biotechnology and stem-cell research in particular. As we shall discuss in Chapter 3, Blair has championed Britain as a world-leader in biotechnological innovation and become a spokesperson for the promise of these technologies nationally (for the UK) and internationally.

As we have argued above, most of the supporters of therapeutic cloning have been vigilant and vocal in their efforts to distinguish such applications from reproductive cloning. In addition, in the UK, the Science Media Centre, established in 2002, has played a crucial role in articulating this distinction and in co-ordinating the voices advocating it.[24] This Centre has framed its interventions and those of the scientists whose voices it endorses predominantly in terms of good science and good governance.

However, there is another range of global cloning actors who have generally been characterised in the UK and some other Western media as much more dubious. Richard Seed, Severino Antinori and Panos Zavos and Brigitte Boisselier of Clonaid (the reproductive cloning enterprise established by the Raelians), have each made widely publicised claims about their human reproductive cloning endeavours, none of which have been supported by credible evidence.[25] Nonetheless, media interest in cloning means that, much to the chagrin of the scientific establishment (Haran 2007), these actors share the same discursive terrain as Ian Wilmut, Stephen Minger and the HFEA. We explore the dimensions of this sharing and its contestation in the following chapters.

The distinctive political configurations in the UK, USA and South Korea mean that discursive actors opposed to, or sceptical about, therapeutic cloning and stem-cell research face very different challenges in getting their messages across in these different national contexts. In *The Embryo Research Debate*, Michael Mulkay documents the extremely effective campaign realised in support of embryo research in the late 1980s in the UK by 'an association called Progress . . . under the joint leadership of scientists, physicians and parliamentarians' (Mulkay 1997: 28).[26] This campaign was successful in establishing the legal principle that research could be conducted on human embryos until they were fourteen days old, which led to the 1990 legislation we outlined above. Beyond this, it also effectively positioned opponents of such experimentation, who did not share this scientific conceptualisation of the embryo, as a separate moral community that was considered to be outside the democratic, scientific and secular mainstream, and therefore incompetent to intervene in any future debates about embryo research. This strategy continues to be effective in the present conjuncture, as few UK-based science journalists are prepared to take religiously motivated objections to embryo research seriously. Nevertheless, despite this negative framing, anti-abortion organisations, which have expanded their activism to encompass critiques of embryonic research, such as Comment on Reproductive Ethics (CORE) and LIFE, continue to be active in UK public discourse.

In the USA, despite the bipartisan framing of the debate mentioned above, one pressure group, the Center for Genetics and Society, has been extremely active in presenting critical perspectives on genetic technologies, including embryonic stem-cell research and therapeutic cloning.[27] This Center is also pro-choice (that is, in favour of women's right to legal abortions) and has been instrumental in launching legal challenges against Proposition 71 and against the interventions of the California Institute for Regenerative Medicine.

In South Korea, critical voices come from civil societies concerned with democracy, environmental issues and women's rights. The network 'Women-Link' has been particularly effective in raising concerns. However, a strong sense of the link between patriotism and support for Hwang's work within South Korea made open criticism difficult for many years.[28]

Conclusion

This chapter has explored the changing meanings and cultural imaginings of cloning. One strand of this has been our critical reading of various genealogies that have emerged recently, particularly in the wake of announcements of the cloning of Dolly the sheep (1996) and of the completion of the Human Genome Project (2000). This reading suggests that there are no disinterested versions of this story and we have expressed some scepticism about any origin story or single 'master' narrative. Instead we offer a brief outline of a set of genealogies which we think are all important in the history of the recent cultural imaginary of cloning. These include: the history of experimentation in animal embryology; the developments in new human reproduction, particularly IVF in the late twentieth century; the trajectory of excitement and controversy that has shaped the Western cultural imaginary of cloning; the literary and filmic traditions of representing cloning, particularly in the genre of science fiction; and recent examples of naturalising or normalising discourses.

Our interrogation of stories of the history of cloning was also a lever for our second mapping exercise: charting some features of the recent discourse on human cloning. We highlighted the emergence of human cloning as a plausible, imminent technoscientific prospect as a key feature of the recent discourse.[29] The focus on SCNT and SCNR as crucial contemporary versions of cloning and the extensive circulation of visual imagery associated with SCNT and SCNR we contended have also been important aspects of the contemporary Western cultural imaginary of cloning.

The sketching of the recent shifts in the meaning and imagery of cloning included the delineation of two sets of discursive distinctions which loom large in early twenty-first century discussions about cloning: fact and fiction; therapeutic and reproductive. We began exploring the emergence and discursive investment in these distinctions which we will pursue throughout the rest of this book. However, we have also drawn attention to the tensions and instabilities around these.

The final component of our exploration of what cloning is in the early twenty-first century was our brief sketch of some features of the national and international discursive terrain of cloning. Here we introduced some of the key legislation governing cloning in the three national contexts considered in this book: the UK, the USA and South Korea. We included with this a brief outline of initiatives to instigate international regulation in this field. This was supplemented by our introduction to some key discursive actors in each of these settings who have been prominent in recent cloning and related debates and who figure in the analysis which we offer in the following chapters.

This scene-setting chapter used two media events to demarcate the chronological starting point for our investigation: the announcements of the birth of the clone, Dolly the sheep (1997) and of the 'completion' of the Human Genome Project (2000). This chapter has furnished an account of how cloning has acquired a new set of meanings, imagery, and associations in the decade since the media broke this news. In subsequent chapters we tell our version of some of the dimensions of the mediated making and re-making of cloning in the late twentieth and early twenty-first centuries.

3 Cloning futures

Introduction

In this chapter we focus on the ways in which the discourse of cloning has evolved during the period between 1997 and 2007, to produce particular versions of the future. This entails exploring and comparing recent modes of conjuring a positive vision of the scientific future and the repudiating of technoscientific dystopias. In the previous chapter we examined some of the new meanings which have accrued to cloning and how somatic cell nuclear transfer (SCNT) and stem-cell research are currently co-constituted and linked with cloning. As we indicated, prior to the 1990s, human cloning was identified mainly as a trope of science-fiction films and novels. These offered dystopic versions of human reproductive cloning focused on the visual figure of the clone as multiple duplicate as in *The Boys from Brazil* (1978) and *Parts: The Clonus Horror* (1979). However, in recent years human cloning is no longer predominantly associated with this dystopic vision, instead it has been increasingly linked to a promising future of therapeutic technoscientific practice.

In the previous chapter we showed how, in the context of UK technoscientific policy debate, cloning became linked to a 'rhetoric of hope' through its implication with the discursive legacy of the embryo debates of the 1980s (Mulkay 1997). In this chapter we explore the rhetoric of hope further in order to examine the emergence of specific national imaginaries about cloning. We explore how cloning is presented as offering the promise of regeneration and relief from suffering. As regenerative, cloning is re-constituted as a medical *process*, as well as being figured through the *body of the clone*. In this chapter we argue that human reproductive cloning is increasingly figured, in some instances, as a possible future. It is the specific features of the recent constructions of both the promise and fears associated with cloning that we address here.

This chapter highlights how human cloning has come to be identified with certain interwoven features – the prospects of:

1 future cures and freedom from suffering;
2 temporal contraction and imminence;
3 global subject positions.

We structure this chapter by working through these three features in detail. Our emphasis in the last chapter was on the instability of recent cloning discourse. We noted there the long trail of discredited scientific claims and the well-established dystopic tradition associated with science fiction. This makes the recent production of therapeutic cloning as a hopeful future a rather complex operation. As we shall demonstrate, the rendering of therapeutic cloning (by its advocates) as hopeful, in the recent conjuncture, often requires both the repudiation of previous fraudulent claims and the refusal of reproductive cloning. Nevertheless, in very specific contexts (which we will elucidate below) reproductive cloning has also been figured as a hopeful possibility. As we shall demonstrate, in the last few years, such expectant conjurings of reproductive cloning have emerged only in quite distinct arenas. These arenas include: the statements of 'maverick' scientists, or statements addressed to very specific audiences (such as 'select' policy committees), who are regarded as being more rational than others, and statements in very contemporary film. More generally, since 1997, human reproductive cloning has predominantly been associated with horror and dystopian science fiction. It has also been identified with fraud (Kolata 1998), fakery and inauthenticity, the uncanny, the so-called 'yuk' factor[1] and homosexuality (Franklin 1999; Battaglia 2001; Stacey 2003). In recent film and media news coverage, scientists engaged in reproductive cloning have fairly consistently been represented as unsympathetic characters – 'maverick', weak, misled or evil. Although we shall consider some variation over time later in this chapter, it is important to register that only four of the most recent films (out of thirty-five films dealing with genomic themes in the period of our study) represent cloning equivocally or favourably. These equivocal or positive portrayals may be found in: *Blueprint* (2002), *Teknolust* (2002), *Code 46* (2003) and *Aeon Flux* (2005).[2] Historically, reproductive cloning (and hence the figures of the duplicate child or adult clone) has been the focus of film. By contrast, therapeutic cloning (and the figure of the blastocyst or embryo) became the focus of news reporting. However, in recent years there has also been a period of instability and change within cloning discourse. The possibility of the child/adult clone emerges in some news and policy discussion, while, alongside this, the image of the blastocyst/embryo (and accounts of therapeutic cloning) have begun to appear in films and novels. This is particularly true of some very recent cloning films which explore process and therapy as well as reproduction. In the following discussion the shifting appearance of the figures of the clone, the foetus and the embryo is partly what intrigues us.

We also examine the subject positions offered through projections of the promise of cloning, particularly the 'universal' human subject associated with cloning cures. In so doing we are mindful of Donna Haraway's observation that: 'Technoscience is the story of . . . globalisation; it is the travelogue of distributed, heterogeneous, linked sociotechnical circulations that craft the world as a net called global' (Haraway 1997: 12). Thus, we look at how the human subject identified with therapeutic cloning emerges through narratives about national identity to appear as a globally located national subject, at once both local and global. We endorse Haraway's argument by showing that this specific version of the subject does not appear as universal, but rather, is constituted through contemporary technosciences as global (Haraway 1997; Franklin et al. 2000).

Future cures and freedom from suffering

In this section we pay attention to the way that the meaning of cloning is constituted as freedom from suffering in the future. Between 1997 and 2007, SCNT and the development of DNA-matched stem-cell lines have been promoted as one of the most promising processes through which scientific medicine will be able to cure disease. There has been some unevenness to this, and some variation, with attention sometimes being focused on DNA matching and at other times on making embryos available for research. However, the foregrounding of SCNT as the most promising technoscientific process for realising medical 'cures' has become a recurring and reiterated rhetorical position in some quarters. This was somewhat disrupted by the Hwang scandal that emerged at the end of 2005. However, it was not entirely undermined. For example, in the Playfair Lecture delivered at the University of Edinburgh, 19 January 2006, Ian Wilmut laid out the scientific steps, imagined as the path to stem-cell cures for motor neurone disease by using a schema in which an illustration of Hwang's work appeared but had also been crossed out. Despite the controversy and uncertainty surrounding such research at that point (particularly in the wake of Hwang's discrediting), the process of SCNT was still clearly identified as facilitating the path to cures. Wilmut's proposal extended to the creation of hybrid rabbit/human embryos for experimentation. Even at a time when the project of SCNT embryo production and the matching of stem-cell lines was coming under scrutiny and even in the context of a controversial and novel proposal, Wilmut reasserted that cloning *would* lead to cures by presenting his work as a feasible and promising pathway for curing motor neurone disease.

In the UK, the USA and South Korea, the message that stem-cell cures constitute the apex of contemporary biomedical achievement derives, in large part, from the recent conjurings of embryonic stem cells as magically pluripotent. These expectations about the magical pluripotency of stem cells and the linear progress narrative of future hope associated with these expectations has been graphically figured in the South Korean stamp (2005) which commemorated Hwang's work (see Figure 3.1).

This stamp visually encapsulates the recent expectations for embryonic stem cells and for their potential in providing future freedom from suffering. It presents a procession of images: starting with enucleation and ending with the human figure who is encoded as male (through trousers) leaping from a wheelchair into an embrace with a figure who is encoded as female (through the skirt). The miraculous power attributed to the embryonic stem cell is evident in this image which resonates with the image of Jesus Christ causing

Figure 3.1 South Korean stamp, an example of visual material suggesting swift and miraculous stem-cell cures as the outcome of SCNT cloning

'the lame to walk'. Such power, which has traditionally been associated with divine intervention, has been discursively associated with recent developments in the medical biosciences and figured specifically through the embryonic stem cell. The image of the figure leaping from the wheelchair is visually overshadowed by, and emerges out of, the image of SCNT or therapeutic cloning. Thus, therapeutic cloning occupies centre stage in the vision of the future of regenerative medicine, reiterating and remaking the performative promise of embryos and stem cells (Mulkay 1997; Parry 2003; Williams et al. 2003; Franklin 2005).

Temporal contraction and the imminence and proliferation of cures

In this section we examine the ways in which human cloning and embryonic stem-cell cures are imagined as currently available or imminent (Kitzinger and Williams 2005). The South Korean stamp is a limited example of visual material, but it provides a point of departure for some of our key arguments. This temporal contraction can be seen in the image on the stamp which constructs the sequence from cell enucleation and transfer to cure as smooth, inevitable and quick. This suggests miraculous cures as the outcome of SCNT. It invites the viewer to assume that, since cloning technologies (such as that depicted) are available, technoscientific cures are also imminent. The repetition across multiple genres of the connection between immediacy, cures and cloning constitutes a branding of cloning through stories of biomedical hope.

This striking stamp is not unique. Indeed, it is emblematic of a more pervasive pattern of discursive renderings involving the condensation of the timelines for curing through technoscientific cloning in recent news reporting and popular science writing. More precisely, it has become conventional to invoke a timeline of five to ten years in commentaries about the promise of genomic medicine. The phrasing 'five to ten years' is infinitely repeatable; because no definite date is given then it has no expiry date. However, this particular timeframe (five to ten years) emerged before therapeutic cloning had been realised and has hence been in operation for more than ten years. Of course, the reiteration of such rhetorical tropes (five to ten years) would seem to require continuous adjustment and readjustment of expectations. It could be argued that, in some ways, this is what has been going on in the extensive coverage of genomic science and medicine since 1997. However, the conventions of scientific journals and news reporting and values can obscure such discrepancies and the rhetorical nature of such timelines, since media news coverage generally does not register announcements about unsuccessful medical trials or the failure to realise a cure. Some of our audience research showed that there were high expectations that cloning would yield cures imminently and, indeed, that it had already done so. In fact, in response to the question 'How do you know about genes, genetics and cloning?' respondents identified a range of innovative treatments which they attributed to technoscientific cloning. One respondent to our Mass Observation directive wrote:[3]

> A recent step in the cloning field that I found amazing was when they cloned a new face for a woman who had been disfigured by a dog bite. Good luck to her.
>
> (MO E3624)

The association of this face transplant (which received international media attention in 2005) with cloning techniques also featured in several of our focus groups. In fact, the face transplant did not involve cloning – understood as SCNT – at all. Neither was cloning involved in many of the other types of treatments mentioned by our research participants.[4] People spoke of cloning being responsible for hand transplants and growing bladders and other forms of tissue engineering such as that associated with the widely circulated image of a human ear growing on the back of a mouse.[5] Such comments illustrate and register the condensation of the timeframe for medical innovations associated with cloning and the general sense that cures deriving from therapeutic cloning are not only imminent, but have already been realised.

The impression that there is a proliferation of therapeutic cloning cures is created through the use of juxtaposition and reiteration in making bids for audience attention. Creating such connections and maintaining their intelligibility across texts and sites involves, in Celia Lury's terms, the creation of a brand (Lury 2000). Research suggests that simply reiterating associations may have a powerful impact on how people understand an issue, or a product. For example, media accounts often juxtapose discussion of the hole in the ozone with discussions of global warming. Public surveys show that this has left many people with the (incorrect) impression that the former *causes* the latter (Hargreaves, Lewis, and Speers, 2003). In fact such associations, frames and branding may be more significant in shaping reactions to an issue than publicising any number of facts (Kitzinger 2007). This is clearly evident in public discussions of a range of emerging biotechnologies. GM crops, for example, have been rejected in the UK. Focus group participants reject the GM enterprise as 'industry led' and about money, rather than associating it with consumer well-being. GM food is merged in people's minds with fast food, chemicals and preservatives. They may not know what genetic modification of plants involves, but they associate it with 'the bad'. By contrast, therapeutic cloning is generally part of a discourse of medical progress (albeit disrupted sometimes by reproductive cloning fears). Mention stem-cell research/ therapeutic cloning and many people will immediately start talking about medical benefits in ways which feed on, and feed into, positive expectations (Hughes, et al. 2006).

The cloning scientists we interviewed were reflective about the problems caused by the temporal contractions around their work. Sometimes they attributed the problem simply to media bias: journalists misrepresenting statements by scientists. Ian Wilmut, for example, complained that because of a 'sound bite culture, it [the media] tends to be dealing with very small issues very quickly and [has] a tendency just to sensationalise it, and a tendency to look for an outcome within a very finite period of time, which can be misleading' (interview with authors).[6] At other times the problem was identified as an over-eager public: audiences misinterpreting statements by scientists. Alison Murdoch, for example, commented on how she thought people had understood her statement to news reporters when the Newcastle team were granted a licence to create a cloned human embryo in 2004:

> Well I can remember the first time we got the licence through [to clone a human embryo] and I went on the *Today* programme [a flagship BBC radio news programme in the UK] and was saying . . . I remember the sentence clearly, and I said 'In ten years time we hope that this will happen to this child who has got diabetes.' And I think everyone picked up the 'child with diabetes' and no one listened to the 'in ten years'. But I said it! I don't quite know how you get over that because people hear what they want to hear. And . . . all you can do is keep on saying it, I know that . . . scientists have been challenged for perhaps over-egging the potential, as they say, [but] I don't see that, I think it's just the way people pick it up, people are perceiving it.
>
> (Alison Murdoch, interview with authors)

Alison Murdoch also, however, suggested that a more complex tension might contribute to the problem. Scientists, such as herself, do not want to create false expectations, but also may feel under pressure to promote the potential of their research in order to attract funding:

> I have a problem with patients who phone up and say 'Can we join the clinical trial?' Now that, obviously, is a problem because I have to say to them [. . .] 'I appreciate the problem you've got now, but you are not going to benefit from this.' So from that point of view it's difficult. [On the

> other side] . . . we have the problem that we need to get funding for these studies, and part of funding requires us to say 'Well, what's the potential benefit of this?' So when we put an application in to the MRC or whatever it is to fund a project, we have to say 'this has got huge potential'. And how else do you say it? Do you say 'Well, maybe [in] ten years time it might benefit', and someone will come up with a super-drug that they think is going to work in two years time, the money goes to the super-drug in two years time. So, we're slightly in a catch 22, speaking to different audiences at different times.
>
> (Alison Murdoch, interview with authors)

Other factors also come into play in creating this climate of expectation about imminent cures, not least the conventions around news generation and their relationship to scientific publishing. While recent fraud cases have been crises which had to be managed (see Chapter 4), the news media, such as scientific journals, offers few announcements that a certain technology has not lived up to its original promise. In this way, news reporting tends to reiterate and amplify technoscientific promise. Hence, in the recent news coverage of embryonic stem-cell research in the UK (and probably elsewhere), there has been little commentary about the *lack* of actual cures from embryonic stem-cell research. One of our Mass Observation respondents offered the following acute diagnosis of recent media coverage of genetics:

> The news is quite fickle on the subject of genetics, at one point we were inundated with news of Dolly the sheep . . . then we hear all about genetically modified food, then nothing. It's as if they can't be bothered to follow the stories up and we are just left wondering what the implications are for the future.
>
> (MO B3639)

The evaluation offered by this research participant also mobilises a persistent form of argument that we have identified in discussions of recent developments in the biosciences. This involves the attribution of *any* problems emerging in this field to the media rather than to science or to any social or political agents or agency. (We develop this point further in Chapter 7 when we discuss the constitution of publics and audiences.)

This pattern of temporal contraction and the conjuring of cures as imminent and prolific is also instantiated through the specific reporting conventions which have developed in the coverage of developments in genomic science. For example, when the production of a SCNT embryo in South Korea first made the news in 2004 it was linked to cures for cancer, diabetes, spinal injuries, Alzheimer's and Parkinson's diseases and to the generation of replacement organs. When the announcement of the HFEA licence for therapeutic cloning was made (also in 2004), it was reported with commentary about the 'potential' of stem cells. This announcement became an occasion for the licence holders,

Alison Murdoch and Miodrag Stokjovic, the HFEA and many UK journalists to reiterate that therapeutic cloning was about 'cures' not about 'playing God'. An example of this is the *Scotland on Sunday* article by Katie Foster 'Hopes and Fears as UK Sends in the Clones' (15 August 2004), in which the discussion of the potential for cures takes up the main part of the article. However, this dominant theme was also set against a quotation from LIFE spokesperson, Dr Jack Scarisbrick who referred to 'playing God' and 'cloned babies'.[7] The issue of whether or not cures will be realised is not a matter for debate. Instead the structure of the article sets up a different opposition. The assurance that the future of cloning in the UK was to be realised as *therapeutic*, rather than *reproductive* cloning, contains the fears associated with the latter, while also reiterating claims about therapeutic potential without any contestation. The future of cloning as exclusively therapeutic was thus proclaimed through the evocative disavowal of 'playing God', which refers, in fact, to undertaking reproductive cloning.

Stem-cell cures have thus been constituted as certain and imminent in recent discourses of cloning which affect various forms of temporal contraction. Returning to the image of the South Korean stamp considered in Figure 3.1, the person is depicted as leaping from his wheelchair in the images set on the right, as a result of research involving therapeutic cloning, illustrated through the image on the left. Western reading conventions mean that this stamp is likely to be read as a progress narrative in the literal sense. The use of different colours and lines between the frame containing the cloning image and that containing the restored patient denotes a temporal sequence. Nevertheless, the enclosure of images denoting both research and cures within the single frame of the stamp itself evokes a temporal contraction between the present and the future, between currently available SCNT (of the illustration on the left) and the miraculous cure (shown on the right). Hence, SCNT is unambiguously represented as curative. The inevitability of cloning being curative is further emphasised by the evolutionary connotations of the imagery. This echoes racialised schematic representations of human development, which had considerable currency in the 1970s (for example, in Jacob Bronowski's, *The Ascent of Man*, 1976), picturing human evolution through a sequence of images – from the crouching primate – to the upstanding 'man'.

The distinctive temporality and the condensation of the timeframe for the realisation of cures through therapeutic cloning have been powerful features of recent representations of this field. As Ian Wilmut's Playfair lecture, cited previously, illustrates, the promise of cures can be mobilised in attempts to attract funding and to justify controversial experimental scientific research. The expectation that therapeutic cloning is providing or will imminently provide cures also underpins the proposition that opposing stem-cell research or therapeutic cloning involves a deliberate rejection of (the possibility of) alleviating suffering and curing the sick. Ian Wilmut's address to the Genomics Forum in Edinburgh (22 September 2005) repeated this proposition. In this instance he invited the audience to imagine that they (or people they knew) had motor neurone disease in order to make the point that his research was linked to imminent cures. Such arguments have had notable currency amongst some scientists, but they have also been aired by some politicians and bioethicists (Harris 2004).

While the news media have generally presented the possibility of cloning becoming successfully therapeutic as an imminent prospect, they have tended to repudiate reproductive cloning. The future of reproductive cloning has fairly consistently been associated with fictional dystopias and, as we show in Chapter 4, with maverick scientists. There has, however, been greater variation in film treatment of reproductive cloning over the last few decades. In fact, such representations have shifted from what had appeared to be a relatively stable mode of representing reproductive cloning through dystopian narratives of mass production (*The Boys from Brazil* (1978), *Parts: The Clonus Horror* (1979), *Blade II* (2002), *Star Wars Episode II: Attack of the Clones* (2002)) to more recent tales of individual restoration through reproductive cloning (*The Island* (2005), *Aeon Flux* (2005)).

In many of the films distributed since 2000 the temporal contraction, traced in other media forms, is also apparent, but this trope has been employed to highlight recent problematic developments in the life sciences, especially the commercialisation of human life. So, for example, *The 6th Day, Code 46* and *The Island* present visions of the instrumentalisation of the human body through bioeconomic scenarios, including the selling of tissues. Although the scenarios of these films are construed as futuristic, they all conjure a 'near' future and they make references to contemporary issues pertaining to cloning. Although *The 6th Day* and *The Island* differ from *Code 46* in that they are big-budget, mass-marketed and effects-led action films, all three are aesthetically realist films which visually evoke a present/future world. They also revolve around actual dates, events or consumer goods that contemporary audiences can identify as familiar elements of their own world. All three of these film narratives exist in an imagined space in which reproductive cloning is technically feasible and extensively employed. In this respect, they occupy a narrative space which, as we argued in the last chapter, was opened up by the birth of Dolly the sheep, which became both the phenomenological and symbolic evidence of the possibility of human reproductive cloning. This is made visually explicit in *The 6th Day*, which uses images from the news reporting on Dolly as part of the opening credit sequence. This film thus echoes the commentaries and policy responses of 1997 by reading and reproducing Dolly as the signal of the initiation of human reproductive cloning.

Reproductive cloning has been the subject of international policy discourses since the announcement of Dolly's birth. More recently, reproductive cloning has been the focus of specific discussions in policy discourse in the UK in relation to reproductive technologies. The *Report of the UK House of Commons Select Committee on Reproductive Technologies and the Law* (Fifth Report), which appeared in 2005, precipitated a more optimistic approach to human reproductive cloning in the media.[8] Whereas media news coverage in the UK, as we have indicated, fairly consistently repudiated reproductive cloning, this Report appeared to entertain the possibility of reproductive cloning as a feasible scientific practice. This was construed through references to arguments made by Ian Wilmut and to public attitudes research conducted by Sarah Parry.[9] The Report thus provided a point of entry for the (controversial) idea that reproductive cloning could become a legitimate technoscientific practice in subsequent UK, and global, news reports. The Report itself made the case that global regulation was patchy and that it was therefore likely that someone would try reproductive cloning at some stage somewhere else (other than the UK). It also argued that the only convincing case against reproductive cloning was the dangers it posed for those on whom the experiments would be carried out (see Chapter 4 for an examination of how women's bodies appear and disappear in these debates). Giving evidence to this Select Committee and drawing on her work on public attitudes in the UK, Parry reported that her research respondents did not expect reproductive cloning to remain a tabooed practice. Thus, in this very specific context, an optimistic view of the future prospects of reproductive human cloning emerged.

Our research offers further evidence of this culture of expectation in relation to reproductive cloning in the UK. Our interviews with members of public, and scientists, our Mass Observation consultation, and focus groups accessed diverse expressions of opinion towards reproductive human cloning. Our respondents to the Mass Observation directive presented a range of views, with some being equivocal and a significant minority finding this technoscientific practice abhorrent, but many indicated that they regarded the prospect of the use of reproductive cloning as acceptable. Indeed, a group of school children with whom we conducted a workshop in 2006[10] articulated the assumption that there was nothing particularly novel about reproductive cloning and some thought that it probably had already been carried out. Some overlapping findings emerged from our survey conducted at a film festival on the ethics of human cloning (Edinburgh Film House, 11–13 November 2005).[11] This research indicated that many participants thought that human reproductive cloning was feasible and possibly already operationalised, although they were divided about whether they thought this was an abhorrent or acceptable prospect.

Similar comments were echoed in some of the focus groups. Although most 'drew the line' at reproductive cloning, a few participants thought it might be acceptable:

> I can kind of understand it. … Say you tragically lose a child to something, if somebody actually turned round to you and said 'I can give you a chance to have your child back, exactly the way it is'.
>
> (Imogen, Group 12)[12]

> Kevin: The actual idea in there (*The Boys from Brazil*) was probably okay in my eyes, it was just the way he went about it was wrong.

Interviewer: So it didn't make you anti the idea of cloning?
Kevin: No, it was just the person that he was trying to clone. If he was trying to clone a good person, maybe, you'd look at a different angle.
Interviewer: So if you take a film like *The Boys from Brazil*, might that ever become reality?
Betty: Yeah, I think it could.
Interviewer: Does it worry you?
Kevin: No.

(Group 11)

These comments indicate that in the last few years, in the UK at least, cloning has been normalised as a practice associated with curative and regenerative medicine and that *reproductive* cloning is becoming a potentially acceptable and imaginable practice, at least among some people.

The appearance of a plethora of popular science books on cloning has contributed to this normalisation of human cloning. It is the life blood for the booming market for popular books, including the offerings of the authors we considered in Chapter 2 (Ian Wilmut, Gina Kolata, John Harris and Arlene Klotzko) as well as a host of other popular science writers such as Lori Andrews, Armand Leroi and Stephen E. Levick. Such popular sciences books are complemented by novels by Kazuo Ishiguru, David Mitchell, Margaret Atwood and Michel Houellebecq, which have also contributed to an imaginary in which human reproductive cloning could soon become a normal techno-scientific practice. This sense of the imminence of reproductive cloning has also influenced other channels of cultural production and resulted in cloning scenarios in films and television drama, as well as news media, together with representations of cloning in the arts (Anker and Lindee 2003: 124–31). This configuring of cloning on a linear grid and the condensation of timescales to render cloning and its benefits inevitable emerged in different ways in different contexts.

As we have indicated, the repudiation of *reproductive* cloning operates to legitimise and secure investment in the present and future of *therapeutic* cloning. However, we have also emphasised that this discursive distinction is highly unstable and, as indicated above, there have been openings within cloning discourse, which suggest the possibility of a promising future for reproductive cloning. Moreover, the proliferation of cloning narratives which are linked to cures and the alignment of cloning with IVF has undermined the established and exclusive associations of reproductive cloning with horror and science fiction (see Chapter 2).

Indeed, a notable feature of recent reproductive cloning narratives in film is that they are no longer confined to horror or science fiction. These have also filtered into mass-market action films (*The Island* 2005), and niche market/independent productions (*Blueprint* 2003, *Aeon Flux* 2005). In these films reproductive cloning has been constituted as a technique which provides cures for individuals or the ultimate in DNA-matched organs and tissue. Cloning also appears as a form of assisted reproduction and as a technology which saves humanity from extinction. Thus, cloning, as a technoscientific mechanism identified with the regeneration of existing bodies and the generation of future bodies, becomes part of these visions of remaking human reproduction.

So far we have examined how SCNT has been linked to freedom from suffering and cures, and how this is constituted through temporal contraction and branding. These associations have also reconfigured understandings that cloning is horrific, to contribute to a cloning imaginary which is largely structured through hope, although fears also play across this. In the next section we move on to examine the third of the three features of cloning futures, the association of the future global subject with cloning.

The future for the global subject

The stories about therapeutic cloning cures which have proliferated during recent years often address a global human subject who may benefit from the anticipated cures which are potentially (although conditionally) available to everyone. In policy and news reports this occurs through the references to 'our future', future generations, and patient group members who will be cured in the future. For example, the UK Department of Health White Paper, *Our Inheritance, Our Future: Realising the Potential of Genetics in the NHS* (2003) is another example of the performative configuring in recent genomics discourses. The title of the paper and the £50 million implementation fund it mobilised illustrate and enact such performativity. The title – *Our Inheritance, Our Future* – portrays genetics as a continuum with a definite future. The 'potential' of genetics is simultaneously both opened up and immediately secured discursively and through the allocation of funding. The use of the word 'our' invokes a subject that is at once part of the British nation and a global subject.

The global subject is also pursued through a focus on individual celebrity patients who will be cured or scientists who will develop cures. Cases in point would include the pairing of Ian Wilmut (who proposed to conduct research into motor neurone disease) with Jimmy Johnstone (the late Scottish former football player who suffered from it), and the pairing of Hwang Woo-Suk with the paralysed rock star Kang Won-Rae. Such pairings have featured in the UK and the South Korean media respectively – and are framed through particular national imaginings. Transnational products such as Hollywood films, also present stories about the global human subject. As already briefly outlined, in film there has been a shift from dystopian narratives about mass production and a focus on the figure of the clone, to stories about individual restoration. In some recent films there is even an emerging narrative, showing salvation through human reproductive cloning (*Aeon Flux* 2005). These various stories about individual restoration through new genomic science and medicine pertain to a global human subject who stands in for humanity in general.

News stories, popular science writing and celebrity biographies and autobiographies about heritage, identity and the elimination of human suffering draw on the specificity of disease and of national context to evoke this potentially global subject. Amongst the most prominent stories of celebrities seeking cures through genomic science and medicine have been those of Michael J. Fox and Christopher Reeve in the USA. These celebrities have been represented as, at once, potential individual recipients of cures and as representatives of their societies more generally. In each instance, their celebrity status and illness evoke both ordinariness and exception. In the context of the USA, assumptions about access to social mobility add a further dimension to the enchantment with celebrities. Michael J. Fox's autobiography is entitled *Lucky Man* (2003), emphasising his ordinariness rather than his special status. One of the strap lines of the Michael J. Fox Foundation for Parkinson's Research is a quote attributed to him: 'in my 50's I'll be dancing at my children's weddings. And mine will be just one of millions of happy stories'. These celebrity figures have developed strong links with patient groups, on whose behalf they attract media attention and personal identification. Jimmy Johnstone (the UK footballer) and Won Rae Kang (the South Korean pop star) are other examples of such stem-cell celebrities. Hence, the figure of the damaged human who is to be cured through stem-cell developments has become a familiar character across a range of media genres. In the UK this has included the television drama *Learning to Love the Grey* (BBC, 2000), the drama documentary *If* (BBC, 2004) (see Chapter 6), the documentary on cloning and cures for disease *How to Build a Human* (BBC, 2002), as well as news reporting on developments in stem-cell research.

In Chapter 2 we briefly discussed the emergence of utopian visions of cloning in the 1960s and 1970s. We also noted the more dystopian framings of cloning associated with much twentieth-century science fiction (literature and film) and with the distillation of fears regarding technoscientific development in the 1990s (see Chapter 1, Kolata 1998). Between 1997 and 2007 there seems to have been some shift in expectations for cloning as these are increasingly focused on the individual and revolve around individual restoration and triumph, rather than collective achievement or transcendence. This is perhaps not surprising given the prominence and purchase of neo-liberal values in the Western world in the late twentieth and early twenty-first centuries. Moreover, such visions resonate with the expectation, in the market-led political contexts of the UK and the USA, that individual choice and agency will be the main mechanisms for the governance and regulation of cloning technologies (Kerr and Cunningham-Burley 2000; Rose 2001; Duster 2003; Heath et al. 2003).

The news of Christopher Reeve's death in October 2004 was framed through stem-cell research/therapeutic cloning rhetoric. Reeve had been one of a group of celebrities who became therapeutic cloning advocates. In fact, Reeve was a particularly poignant figure. As the former star of the Hollywood *Superman* films (1978–83) his use of a wheelchair and belief in the curative powers of embryonic stem-cell research attracted much attention, not just in the USA, but globally. The media images of the wheelchair-bound Reeve advocating stem-cell research were in many ways similar to that offered on the South Korean stamp discussed earlier. Christopher Reeve figuratively evokes 'everyone' as both potentially wheelchair bound and potentially cured. In 2000 Reeve appeared in an advertisement for Nuveen Investments in which, through the use of computer-animated graphics, he was represented walking. The 'rhetoric of hope' (Mulkay 1997) associated with therapeutic cloning and stem-cell research sustained this image of a corporeally restored Reeve. In effect, the computer transformation enacted the promise of these biotechnological processes, producing Reeve as an able-bodied person. In retrospect, the poignancy of this image was accentuated not just by his death, but particularly by its cause, since Reeve died from a very mundane side-effect of his condition – infection from a pressure wound.

As previously noted, in November 2004, 'Proposition 71' was passed in California, backed by the state governor Arnold Schwarzenegger. This Proposition permitted state funding and support for stem-cell research, thereby circumventing the federal ban placed on embryonic stem-cell research by President George Bush (see Chapter 2). Schwarzenegger had appeared in the Hollywood film *The 6th Day* in 2000 as a modernist hero who championed natural modes of reproduction. In this film's 'near future' scenario, human reproductive cloning had been banned and Schwarzenegger's character (Adam Gibson) advocates this policy. Moreover, Schwarzenegger carried this assessment into his contemporary social and political arena by endorsing the prohibition of human reproductive cloning in interviews at the time of the film's release (these interviews were included as extras on the video and DVD release in 2001). Hence, it was through these multiple resonances that the symbolic *Terminator* appeared in the USA political arena in 2004 as an advocate of therapeutic cloning. These oscillations between fact and fiction associated with the personae of Christopher Reeve and Arnold Schwarzenegger have shaped and added to the complexity of recent cloning discourse in the USA and more globally. They also highlight the importance of looking at fact/fiction boundaries and the discursive intersections between (Hollywood) film and news events to interpret these developments. We undertake such an exploration in Chapter 6.

Utopian/dystopian futures are also constituted through the institutions and spectacles of Hollywood film. A technoscientific apparatus which draws on technoscience as a key theme (Cartwright 1995; Wood 2002; Stern 2004), film produces the imaginative space of utopia or dystopia through visualisation. The Hollywood 'dream machine' thereby generates the future and interpellates global audiences into the spectacular power of both film and technoscience and their capacity to create life and to bring futures into the present. Hollywood science-fiction futures are either located in an alternative reality (either temporally or spatially), such as the dystopia of *The Matrix Trilogy* (1999–2003), or in the imminent future which has clear continuities with the present. Recent cloning films generally fall into the latter category. In these films, the dystopian future scenarios linked to human reproductive cloning are shown to be very like the contemporary world and many of its features are continuous with those of the present day (see: *The 6th Day* (2000), *Godsend* (2004), *The Island* (2005)).

The Island (2005) offers a vision of a world in which unlimited healthcare is available through reproductive cloning. The film is set in the USA and shows human bodies being cloned in a secret facility in the desert, contracted by the military but run by a private corporation. The cloned bodies are produced in order to provide body parts, cures and babies for their 'sponsors' from whom the clones are grown. The film references *Parts: The Clonus Horror* (1973) and the plot is similar to the narrative developed by Kazuo Ishiguru in his novel, *Never Let Me Go* (2005). The conceit of *The Island* is that the sponsors believe that the clones are non-sentient 'agnates' (the term used in the film) that exist in a permanently vegetative state. Health is represented as the ultimate consumer good and it is implied that the rich are willing to invest heavily in their efforts to achieve good health, longevity and children. It appears that no one cares to look too closely at the methods they employ to meet their goals. In this film the boundaries between therapeutic and reproductive cloning are blurred and ultimately collapse as reproductive cloning is used as a therapeutic technology and these 'therapies' extend to assisted reproduction when the clones are used to produce babies for infertile clients.

The recent dystopian filmic visions of cloning (in a future which is portrayed as remarkably like the present) also differ from earlier film cloning dystopias in their emphasis on the individual and on individual health care. *The 6th Day, Code 46, Blueprint* and *The Island* all focus on the individual and on cloning as a solution to individual problems, a matter of individual choices and the means for realising personal desires. This constitutes a shift from novels and earlier film images of clones as mass produced and as the result of 'Nazi science', as in *The Boys from Brazil* and from narratives of top–down social programming through biotechnology as in *Brave New World*. Nevertheless, although the newer imagery predominates, older imagery occasionally still appears. For example, the visual iconography of *The 6th Day* evokes mass production, and similar visual coding can also be identified in *Code 46* and *The Island*.

The 6th Day is fairly unambiguous in its condemnation of reproductive cloning through the moral narrative stance of the central character (played by Schwarzenegger) and the representation of those involved in cloning as corrupt, weak or evil. However, it differs from more horror-orientated cloning narratives such as *Blade II* and *Godsend* in that it represents the cloned body as 'normal'. Indeed, the clone is not portrayed as abominable or horrific, but instead appears as heroic. In this and many other respects *The 6th Day* is not a film which frames cloning as substantively horrific. It is the horrors of unregulated capitalism and its harnessing of cloning in morally reprehensible ways which are the main preoccupation of this film.

Code 46, Blueprint and *The Island* are all even more ambiguous than *The 6th Day* in their representations of cloning. In fact, in reviewing the sequence of these films, it is striking that the later the release date, the more ambiguous the film narratives are in their representation of cloning. Hence, while *Code 46, Blueprint* and *The Island* all offer critical portrayals of cloning, their framings are more complex and less straightforwardly negative than those found in earlier cloning films. Cloning is no longer treated exclusively as a horrific practice as it was in the 1970s films, *The Boys from Brazil* and *Parts: The Clonus Horror*. Moreover, as a film topic, cloning is no longer confined to the genres of science-fiction and horror films. Human cloning now features in mass-market adventure films.

As we have indicated, the mass production imagery associated with earlier representations of cloning is redeployed in these recent films. In *The Island* and *The 6th Day* there are shots of multiple human bodies suspended in liquid in what appears to be a pre-conscious embryonic or foetal state. In these films the bodies are adult in form. In *The 6th Day* these are destined to become the repositories – the hosts – for cloned identities which will be transferred to them. Such identities are assumed to consist of the stored optical memories of a person, together with a record of their nervous system. Once the originating 'identity' is transferred, the 'blanks' assume the physiognomy of the person from whom the memories and feelings have been collected. In *The Island* these adult forms are generated from cells, and the growth process from cells to adult form is speeded up to get around the temporal difference between clone generations and to create a cloned body of a living client. These clones again assume the same physiognomy as the individual who was the source of the DNA. Hence, they appear as twins.

However, although the clones in *The Island* are portrayed as originating through the mass production of foetuses, they are clearly represented as individualised. In each of these three films the clones are figured through embryonic and foetal imagery and the cloned bodies are shown growing from cellular bundles into adult form in laboratory-like contexts. However, once the clones are 'born', they are individualised though their characterisation in the development of the narrative. In *The Island* the casting of A-list celebrity actors to play the clones, character development, and the representation of the cloned characters as agents who drive the plot fleshes out the individuality of the clone. Moreover, in this film reproductive cloning is portrayed as a health-care technology available to some individuals. The film thus produces a narrative about good and bad individualism, represented through the individual hero/protagonist clones and the contrasting villain, the capitalist biotechnologist.

In these films, clones are produced for a therapeutic purpose. In this respect, they blur the distinction between therapeutic and reproductive cloning that we have highlighted as a feature of recent cloning discourse and which has been particularly prevalent in news media during the period we have researched. In *The 6th Day* the clones are used to realise health and longevity, so that the identity of the person can be maintained in the event of accident, damage or death to the body. In *The Island* the clones are generated as a source of human body parts, and babies are created to enhance the lives of the clients who 'own' the clones. Thus, in these films the clones represent the promise of enhanced individual health and longevity, re-casting reproduction as therapeutic and therapy as reproductive.

Hopeful and fearful visions of cloning futures

So far we have emphasised hope. In this section we foreground how fears are also present in cloning discourse, even in relation to narratives of cures . Hope and fear do not necessarily come from different actors or sources, and here we examine how fear features as a resource and operates alongside discourses of hope. Stories about the future of cloning are constructed in the mass media by an array of actors including scientists, politicians, NGOs, activist groups, media professionals, bioethicists, social scientists, popular science writers, science communicators and public engagement professionals. However, it is primarily scientists and politicians who have set the agenda by staking the claims for hopeful futures linked to therapeutic cloning in the UK. There has also been some mobilisation of *fears* by some scientists, the Wellcome Trust[13] and the Science Media Centre. Fears, however, are strategically mobilised by these actors in order to enlist support for cloning technoscience, not to challenge it. For example, the claim that Ian Wilmut is not cloning Hitler (Highfield and Wilmut 2006) identifies fears about *reproductive* cloning to make the case for *therapeutic* cloning. In addition, fears about old age, health problems and suffering have been mobilised to encourage hopes for longevity, good health and pain-free experiences. Reproductive cloning appears as the repudiated other in these hopeful future discourses and this disavowal has been reiterated in many sites and in many ways as we have indicated. However, we have also suggested that, more recently, some openings have been created which have made it possible to associate reproductive cloning with such hopeful visions of the future.

However, in the recent period, fears about the future of cloning have also been disavowed in the media when they are construed as in opposition to science. This has occurred through three distinct constructions: first, through presenting fear as the result of misrepresentation; second, by assigning such fears to the marginalised 'other'; and third, by acknowledging fears but presenting them as manageable. The first form of disavowal involves attributing the underlying cause of any misapprehension to misinformation. Such fears are reported by scientists as the effect of 'the public' being misled, usually by science fiction or the media. This argument is based on the assumption that there is no legitimate foundation for such fears and that the public are just responding to negative images of cloning which originate in science fiction or the media:

> Cloning makes us both excited and uneasy because of its sci-fi implications. Perhaps we should not to worry so much. Even if (when) cloning becomes

safe enough to use with humans, it is hard to think of any real demand for it beyond a handful of eccentrics.

(Mike Bygrave, 'False dawns in the brave world of new genetics', *Observer* 22 December 2002)

Such arguments are often sustained through references to dystopic texts such as horror films and novels. *The Boys from Brazil* and *Brave New World* have become benchmark reference points taken as indicative of the power of horror stories which can mislead the public.

These arguments about science fiction undermining public confidence in cloning are somewhat misleading. In fact, one recent study found that pro stem-cell scientists were much more likely than critical activists or members of NGOs to make reference to science fiction in discussions of the public's attitude to such research (Williams et al. 2003).[14] Moreover, as we have indicated previously, science fiction is a complex and ambiguous genre. More specifically we would contend that some recent films do offer hopeful cloning narratives: *Code 46* (2004) and *Aeon Flux* (2005) naturalise or normalise cloning. For example, *Aeon Flux* represents reproductive cloning as the technique which temporarily 'saves' humans as a species from extinction. Furthermore, as we shall demonstrate in Chapter 7, although references to science fiction can be resources which people draw on to articulate the meanings of cloning, science fiction itself is not understood as providing the 'truth' about cloning.

The second way in which scientists and journalists sometimes disavow fears of the future involves attributing such fears to marginal groups who are construed as 'luddite' or extremist (Mulkay 1997). This process is amplified in some news reporting because news values encourage a 'for and against' structure (Kitzinger and Williams 2005). This means, for example, that scientists will be quoted in support of a new scientific development and that efforts will then be made to provide a spokesperson from a group that is known to be 'against' the development. In the UK for example, as we noted in Chapter 2, it has been quite common for Comment on Reproductive Ethics (CORE), an NGO promoting positions which prioritise 'absolute respect for the human embryo', to be placed in this position. Although not all news presentations conform to this formula, it is very rare that dissenting voices are aired in a way which frames such speakers as *scientific* experts. Moreover, this news value structure has not been confined to press, radio and television news reporting, it has also influenced modes of presentation within television documentary and drama in the UK (for example, *If* (2004)) discussed in Chapter 6).

This binary formula of science/anti-science in the reporting of news about science has been complicated by the appearance of 'maverick' scientists in the genomics field. When such scientists have made putatively scientific claims, for example, in announcing that they have conducted reproductive cloning, the codes of practice outlined above are more difficult to apply. In such circumstances, the distinctions between science and science fiction and between scientists and extremists (or 'luddites') are less obvious. Thus, as we will explore in the next chapter, the coverage of Panos Zavos's claims to have conducted reproductive cloning disrupts the standard mode of reporting technoscientific innovation in which hope (about cloning) is construed as rational and orderly, and fears about it as emotional and unruly

In addition to these two constructions which have featured in the reporting of therapeutic cloning in the UK in recent years, there has been a third way in which fears are routinely mobilised, and marginalised, in support of human cloning. This involves presenting fears as manageable. The discursive route here is that the fear is confirmed as natural or normal, but it is contended that members of the public should control or contain such fear and not give in to it. Instinctive fears of the unknown, or gut reactions ('the yuk factor') are acknowledged, but identified as reactions to be overcome through the application of education or rationality. Concerns about abuses of biotechnologies are identified as understandable, but presented as containable. In such discursive constructions, legislation and regulative procedures and structures are cited as mechanisms which should dispel misgivings and reassure the public. Thus, in the UK, references to the HFEA or to the UK's broader regulatory framework are often made in therapeutic cloning stories. Scientists, politicians and journalists thus frequently construe fears of cloning as inappropriate on the grounds that the UK is a well-regulated, 'safe' place for such experimentation (Haran 2007).

Such argumentation may also invoke 'common sense' or appeals for faith in the reasonableness of 'human nature'. Such constructions are particularly evident in various forms of popular science writing. For example, in their recent book, *After Dolly* (2006), Roger Highfield and Ian Wilmut argue that, in regard to the future development of cloning, they/we can have:

> confidence that a well-informed democracy can keep abuses in check; we have confidence in women who would not donate eggs to clone a dictator, but would to help a patient with an urgent clinical need; and above all else we have faith in the vast majority of scientists, who are no different from anyone else in wanting to reduce suffering and make the world a better place.
>
> (Highfield and Wilmut 2006: 244)

In effect, this comment casts those who might object or worry about 'abuses' as misunderstanding or underestimating women and scientists, as well as the capacities of 'well-informed democracies' (such as the UK). Indeed, there is a strange transposition here because, instead of the usual formulation referring to a 'well-informed public', this characteristic is attributed to and constructed as embodied in democracy itself. This seems to underscore the contention that the problem is with those who are ill-informed either about science or about democracy's capacities to regulate it.

As Donna Haraway (1997) has argued, in the late twentieth and early twenty-first centuries, the genome represents the promise of the future. It does this because it offers the prospect that 'we' – humanity as globalised humans

– will have the technological ability to be agents in the reproduction of DNA. However, as Haraway emphasises, the promise of genomics is of the enhancement of human agency within processes that already occur (the reproduction of DNA), so such technological intervention can be construed as 'natural'. Thus, cloning may be naturalised and presented as an already existing form of reproduction at the same time as it is celebrated as a technologised and transformative intervention. *Mission to Mars*, the Raelian doctrine (Rael 2001) and the use of the 'book of life' metaphor in early twenty-first century media coverage of genetic issues (Nelkin and Lindee 1995; van Dijck 1998; Kay 2000) all draw on this seemingly paradoxical framing. From this perspective, cloning is a remarkable technological achievement which should be embraced because we are always already cloned.

Mission to Mars (2000) disassociates cloning from reproduction and represents it as resulting from micro-molecular interactions generated by DNA. The film recuperates and naturalises cloning as a hopeful prospect through an imagined evolutionary history in which DNA was sent to earth by extraterrestrials with whom humans turn out to be kin. In this scenario, DNA operates as a creative force which generates human life and leads to a post-human future involving the reclaiming of extraterrestrial kinship. In contrast, *The Island* (2005) portrays cloning as recognisably a reproductive technology. It provides an anxious, dystopic vision of a scenario which looks remarkably like the contemporary world in which cloning is a technique of the bioeconomy that operates as part of the marketing of human life through capitalist instrumentalism. *The 6th Day* (2000) and *Blade II* (2002) also emphasise the frightening aspects of reproductive cloning.

The aesthetics of these films provide another dimension to their treatment of cloning. In representing the clones as beautiful and by showing health care through cloning as a realisable achievement, *The Island* produces what is, at some level, a promotional image for cloning. This contrasts with *The 6th Day* in which all the clones, with the exception of the main character, are represented as degenerate and in which regeneration is visually linked to horror, violence, gore and blood.

Aeon Flux (2005) represents human reproductive cloning as providing a solution to a global pandemic which threatens to wipe out human life. In this film, cloning is a short-term solution, providing temporal relief from the threat of extinction, but one which must ultimately be rejected with the return to 'nature' at the end of the narrative. Resonant with *Jurassic Park* (1997), which ends with the claim that 'life finds a way', *Aeon Flux* finishes with one of the protagonists declaring that 'nature finds a way'. Hence, *Aeon Flux* concludes with a rejection of cloning on grounds that it is unnatural. This film explores the 'risk society' dilemma that technologies create new problems, even though they have been designed to solve the problems of modernity (Beck 1992). Although it explores the promise of technology (specifically cloning), its resolution of this dilemma is, in effect, to advocate a return to nature (through the rejection of cloning).

Conclusion

In his discussion of utopia and science fiction, Patrick Parrinder posed the question: 'Can we imagine a better society without imagining, and wishing to create, better people?' (Parrinder 1997). His question implies that investments in better futures are co-constituent with investments in normative understandings of fit future subjects. As we have argued, recent utopian visions of cloning as a promising technoscientific mechanism for alleviating human suffering are individualised and figured through narratives of individual choice, rather than state control. Nevertheless, such visions invoke a form of 'flexible eugenics' (Taussig et al. 2003), or a 'backdoor to eugenics' (Duster 2003) because they entail the designation of certain human characteristics as more desirable than others. Moreover, as currently envisaged, while human cloning does not involve the selection of individuals, it does require the selection of body parts, cells, tissues, embryos and foetuses, and investment in the selection of some parts as matter and some parts as human in strategies which mobilise different versions of desirable futures.

Through the image on the South Korean stamp and the figure of celebrities such as Christopher Reeve, the aesthetics of the SCNT-derived able body became metonymic of health. This is crystallised in a scene in *The Island* in which one of the characters explains that the clones represent the 'new American dream' for health and longevity. At this point the (apparent) health and beauty of the cloned characters are in visible contrast with the imperfection of their non-cloned human interlocutor and their immediate environment. Dressed in white exercise clothing and portrayed with glowingly 'perfect' features, complexions and hair, the clones visually represent the technological solution to the problem of human vulnerability to age, accident and disease. Their interlocutor is represented as unattractive and dirty, and as the embodied stereotype of the 'trailer trash' underclass. At the same time this character does seem to represent an interesting and intellectually attractive counterpoint to the bland aesthetic of the clones. Nevertheless, through this stark contrast, *The Island* furnishes a visual representation of idealised, white, Western versions of celebrity, indicating which kinds of bodies are designated as replicable through cloning techniques and which are not.

The utopian vision of a world of enhanced human health and longevity, in which individuals can remake themselves to achieve desirable body traits and become 'ideal' types, is undercut in *The Island* by its focus on the means used to achieve such goals. The film shows that the means employed to achieve this utopia could be very dystopian. Significantly, it is the *social context* and *process of cloning*, rather than the *figure of the clone* which emerges as horrific here. The clones in *The Island* are represented as enslaved beings, denied basic 'human' rights, through the narrative conceit that clones are not human. However, it is always visually explicit to the viewer that the clones are as human as anyone else and the film's rendering of the story world as very like the contemporary USA means that *The Island* tells a reassuring tale of their achievement of freedom. The 'escape of the clones' might sound like the title of a 1950s horror movie, but in *The Island* (2005) the escape is a culmination of the clones' struggle, and a moment which enrols the audience in a sense of pleasure. The fight against social injustice (as experienced by the clones) is led by the 'moral majority' and facilitated by an 'open society', an endless landscape, and the vigilance of the American hero (albeit, played by a Scottish actor). In the euphoric final sequences even the mercenary soldiers turn against the instrumentalism of biotechnology and 'free' the clones/slaves into a version of the unspoilt landscape of the American 'wild west'.

The utopias of current cloning discourses foreground the individual and individual agency and this is a very clear feature of many of the films reviewed in this chapter. In many of these films (see particularly *The 6th Day* and *The Island*) it is the strength of character and moral fibre of individuals which are crucial in the fight against economic instrumentalism and who reassure viewers that technoscience (human cloning) will be reined in and morally regulated. Hence, these films offer a dramatic crystallisation of Highfield and Wilmut's belief in individual good will and heroic agency guaranteeing the good governance of developing biotechnologies, particularly cloning (see Chapter 2).

The potential utility of reproductive cloning is figured positively in *Aeon Flux* (2005). Nevertheless, even in this film which provides the most optimistic reproductive cloning filmic scenario to date (outside of comedy),[15] the biotechnology is ultimately rejected, in favour of a 'return to nature'. The cloned characters, although beautiful, are represented as ghostlike and decadent, disturbed by genetic memories that they do not understand and stretched 'thin' by years of cloning. They are represented as informational simulations: somehow less real than their non-cloned ancestors. This image of the clone draws on the informational model of the genome (van Dijck 1998; Kay 2000; Nerlich et al. 2002). However, despite information providing the capacity for programming perfection (Clinton's 2000 announcement of the 'God's language' of the genome (Nerlich and Hellsten 2004)), a return to the natural is constructed as the more attractive, less superficial and more 'real' option. Thus, in *Aeon Flux*, cloning technologies are not unambiguously affirmed but they do serve a limited function in saving a select group of humans, in providing a temporary future hope and in securing a return to the past.

In news stories about human cloning, the hoped-for future is construed as imminent, whereas, as we have seen, in recent films, the feared future is disavowed and is anyway, by virtue of the filmic form, fictional. As we discussed above, the South Korean stamp which portrays the realisation of cures through the cell nucleation process is a vivid visualisation of the construction of therapeutic imminence – the instantiation of a fused future/present in recent cloning discourse. This temporal condensation is realised through the ritualistic reiteration of terms such as 'hope' and 'potential' in relation to cloning in headlines such as 'Scientists raise hopes for sickle cell patients' (Alok Jha, *Guardian* 10 January 2006), and through the evocation of prospects for epochal changes, particularly revolutions in health care. The ubiquitous rhetorical citing of five to ten year timeframes discussed previously make such achievements seem plausible. In this way, cures become so strongly discursively attached to the prospect of undertaking cloning in the recent UK context that they loom as the inevitable outcome of proposals for experimental research.

The stamp circulates a vision of a utopian future/present of able-bodied humans which is being realised through somatic cell nuclear transfer, which ostensibly enables humans to walk away from disablement and suffering. Cell nuclear transfer operates as the transformative moment – the now – in which present and future are suspended as endless potential and through which hoped-for cures seem imminent. This contraction of time and projection into the promised future has been a feature of recent cloning discourse, but it can also obscure other possible future scenarios.

This examination of visions of the future embedded in recent cloning discourse helps to highlight some of its key features. Expectations for individual health care, fertility, the nation and the global subject have all been telescoped onto expectations for therapeutic (and to a lesser extent, reproductive) cloning. In this sense, recent cloning discourse constitutes this technoscientific practice as multiply regenerative. The following chapters offer more detailed explorations of the contexts through which this is occurring and of the key subjects and objects featured in these discourses.

4 Mavericks, madmen and fallen heroes

Introduction

In this chapter we explore the visual and textual figurings of contemporary life scientists associated with human reproductive or therapeutic cloning. We examine how these scientists are represented in both factual and fictional representations in a range of media genres. We begin by clarifying why the figure of the scientist is so crucial in understanding and investigating modern science. We then enlist the notion of the 'modest witness' (Haraway 1997), reviewing why this critical concept may be useful in exploring the processes whereby the reputable scientist is identified and attributed with 'epistemic authority' (Gieryn 1999). We link this construction of the exemplary scientist with specific processes of the 'virtual witnessing' (Shapin and Schaffer 1989 (1985)) of sciences that are facilitated by the mass media. We then turn our attention to more ambivalent or negative figures whose features contrast with those of the modest witness: the maverick, the madman and the fallen hero. We outline some of the cultural resources drawn upon to frame these figures and we analyse recent versions of these that have emerged in the field of human cloning. The dominant figure of the scientist (whether reputable or maverick) is masculine. However, we conclude by offering a brief analysis of the narrative and visual framing of two women scientists associated with cloning.

It is important to note that the scientific figures mentioned above (the modest witness, the maverick, the madman, the fallen hero) are generally produced, at least implicitly and frequently explicitly, in relation to each other. Furthermore, as Rosalind Haynes (Haynes 1994) and Jon Turney (Turney 1998) have documented, these figures draw on and circulate within a rich stock of historical and fictional resources. In making the distinction between fact and fiction, we are mindful of Haraway's observation that: 'It seems natural, even morally obligatory, to oppose fact and fiction; but their similarities run deep in western culture and language' (Haraway 1989: 3). Not only are we sceptical about such 'natural' or 'morally obligatory' distinctions, but, in this book, we have deliberately set out to trace the circulation of discourses and figures across and between genres and forms that are typically located on both sides of this divide. We develop our main arguments about the interrelationship of truth claims and media genres in Chapter 6. At this point we simply signal our awareness that this is an issue that requires further consideration.

The figure of the scientist

Why is the figure of the scientist important? As we shall illustrate through our specific analyses in this chapter, the figure of the scientist often stands as a synecdoche for the entire field of science. In this sense, the scientist (variously as a specific individual, an ideal type or a stereotype) stands in for the whole of a complex system of practices, institutions, communities and beliefs. Megan Stern argues that there is a distinction between what she calls 'True Science', marked in her text with an upper case 'S', which is 'an idealised concept that is generated by the various discourses and practices which together constitute scientific culture' and 'science', designated with a lower case 's', as it actually exists and manifests itself in the contemporary world (Stern 2004: 347). Stern resists some of the more established versions of what characterises 'public understanding of science' (which we shall discuss later in this chapter), contending instead that: 'public understanding emerges as the site at which an ideal notion of good "Science" emerges, against which the practices of actual scientists often appear inadequate' (p. 348). With Stern's claim in mind, it seems reasonable to posit that negative portrayals of scientists, which seem to cause alarm amongst scientists and policy makers who fear the loss of the public's trust, might serve to establish (by setting up contrasts) what is expected of trustworthy scientists. Hence, rather than demonstrating fear and hysteria, they might be expressions of the legitimate public desire to hold scientists to account. Stern also observes that 'the competing interests and concerns that are usually concealed behind a widely accepted public image of science' become more clearly visible during a period of change in the 'hegemonic currency of science' (p. 350). In fact, the manifestations of competing interests and concerns is perhaps more performative than this formulation would suggest. Such performances of controversy, we argue, may themselves be part of the process of effecting transformation in, or consolidating, the 'hegemonic currency of science'. Our exploration of a range of figures of cloning scientists indicates that the 'hegemonic currency' of cloning science (and stem-cell research) is being forged by scientists and policy makers, and that media figurations of scientists are crucial in that forging.

Images of cloning science are constructed, articulated and contested through the images of the scientists who, in effect, embody this scientific field. Donna Haraway notes that: 'the imaginary and the real figure each other in concrete fact' (Haraway 1997: 2). Hence, meanings of cloning produced through representations of individual scientists, in fictional texts, in which they articulate a range of different positions on and in cloning, resonate with those produced in media discussions of cloning in 'the real world'. José van Dijck argues that, more generally, genetics is best conceived as constituting a 'theatre of representation' (van Dijck 1998: 16). Invoking van Dijck's framework, we are proposing that scientists are one of the key categories of actors in this 'theatre of representation'. Van Dijck's own overview of the figuring of the modern scientist highlights the development and use of the atomic bomb as the most significant turning point in public perceptions and expectations of the scientist. With reference to popular culture, she sketches changes in the dominant image of scientists in this field. She sees earlier twentieth-century versions of the field as associated with a laboratory-based, mysterious and somewhat otherworldly figure who was removed from society (p. 18). She contends that, after the atomic bomb was used, genetics became increasingly identified with scientists who took social responsibility and image management seriously (p. 18).[1] Van Dijck's research also documents a growing awareness amongst genetic scientists about the public image of their profession and it highlights some of the ways they have tried to manage this (for example pp. 68–72).

In addition to the general self-consciousness about image which van Dijck highlights, many scientists are themselves aware of the cultural myths and archetypes that play out in the representation of science. So, for example, in recent news coverage, scientists and some journalists reporting on developments in genomics and cloning, have made reference to stories such as *Frankenstein*. This has usually taken the form of disavowals which are used to indicate misunderstandings of the field or to locate 'other' scientists as 'unscientific' actors. For example, it was almost inevitable that one of the headlines used to announce Panos Zavos's claim to have implanted a cloned human embryo, would be 'Dr Frankenstein' (Thurlbeck 2004). Such practices justify Roslynn Haynes's evaluation that: 'Frankenstein has become an archetype in his own right, universally referred to and providing the dominant image of the scientist in twentieth-century fiction and film' (Haynes 1994: 92). Indeed, as noted previously, Jon Turney has extended this claim, with the contention that the novel from which this archetypal figure emerged 'is the governing myth of modern biology' (Turney 1998: 3).

The modest witness

In the introduction to this book we cited Shapin and Schaffer's (1985) exploration of the social, literary and material technologies that formed the knowledge-making practices of seventeenth-century experimental philosophers as a valuable resource for thinking about more contemporary versions of making scientific knowledge claims and of designating those appropriate to make such claims. The employment and foregrounding of these technologies continues to constitute the (self-)representations of contemporary scientists, who have inherited this legacy of seventeenth-century experimental philosophy. Such technologies are crucial in the constitution of the community of peers and in the practices of boundary management that underwrite scientific credibility (Gieryn 1999).

The modest witness of science was to be 'such a man as should be believed' (Shapin and Schaffer 1989 (1985): 65) and, according to Shapin and Schaffer, the literary technology that Robert Boyle developed was designed to assure that the experimenting natural philosopher would be believed. Thus, Boyle insisted upon a simple, functional writing style, contrasting the modest factual claims characteristic of these early experimental essays with the more grandiose claims of those who set out to write about entire natural philosophical systems. Boyle was explicit that he would only draw tentative conclusions about causes and that he would concentrate instead on the communication of observed and observable facts. Shapin and Schaffer conclude:

> Boyle's 'naked way of writing', his professions and displays of humility, and his exhibition of theoretical innocence all complemented each other in the establishment and the protection of matters of fact. They served to portray the author as a disinterested observer and his accounts as unclouded and undistorted mirrors of nature.
>
> (Shapin and Schaffer 1989 (1985): 69)

Haraway identifies 'self-invisibility' as 'the specifically modern, European, masculine, scientific form of the virtue of modesty' (Haraway 1997: 23), as a form of modesty 'that pays off its practitioners in the coin of epistemological and social power'. Of course, this '[self-]invisibility' does not extend to others. Such modesty is only available to those who possess the appropriate cultural capital and epistemic authority. Hence, such modesty is in marked contrast with the more spectacular characteristics attributed to the other scientific figures we consider here – mavericks, madmen and fallen heroes.

Virtual witnessing

According to Shapin and Schaffer, the witnessing and written accounts of experiments were crucial to, and indeed constituted, the foundation of modern science. The acceptance of experimental accounts as reports of matters of fact depended on the testimony of a signed register of witnesses, whose credibility 'followed the taken-for-granted conventions of that setting for assessing individuals' reliability and trustworthiness: Oxford professors were accounted more reliable witnesses than Oxfordshire peasants' (Shapin and Schaffer 1989 (1985): 58). Those who read the accounts of such witnessed experiments could count themselves as virtual witnesses because the local witnesses could be relied upon to provide accurate testimony. Such witnessing was harnessed to the literary display of functional writing described above. We maintain that this yoking together of the credible witness and an apparently understated, ostensibly transparent way of writing, to produce the figure of the modest witness, is still detectable in contemporary negotiations and contestations about what a good scientist is and should do. Hence, representations of particular contemporary scientists that do not evoke the figure of the modest witness may generate doubt about an individual's scientific status. Some scientists, in some social contexts, do not fit the mould of the modest witness figure, even if they appear to embody some features of the normative culture of science.

Elizabeth Potter challenged Shapin and Shaffer's framework by proposing that a particular kind of masculinity was fashioned – and not simply drawn upon – in the casting of the modest witness in seventeenth-century Britain. Potter delineated how the making of modern science also entailed the making of gender through the production of some men as legitimate and credible witnesses to matters of fact and the barring of other men and all women from this community of witnesses. This approach has been taken up and developed by Donna Haraway (Haraway 1997: 30). Extending Potter's and Haraway's revision of Shapin and Schaffer's interpretation, we argue that particular recent versions of the masculine modest witness have been cast in human genomic technoscience in the UK, the USA and South Korea. Such casting means that women scientists are still not easily registered as legitimate witnesses to scientific truth. As van Dijck points out in her description of the roles allotted to characters in popular accounts of genetics from the twentieth century: 'By virtue of their ascribed feminine characteristics, women hardly fit the category of scientist at all; they are allotted minor roles, and their representation qualitatively differs from their male counterparts' (van Dijck 1998: 24).

In this chapter, we draw on and extend Shapin and Schaffer's notion of 'virtual witnessing' in our analysis of the various configurations of the genomic or cloning scientist: 'the technology of virtual witnessing involves the production in a *reader's* mind of such an image of an experimental scene as obviates the necessity for either direct witnesses or replication' (Shapin and Schaffer 1989 (1985): 60). We argue that the verisimilitude that Boyle attempted to generate through his naked way of writing and the use of expensive engravings to provide images of the experimental apparatus is still pursued in contemporary culture. However, invitations to virtual witnessing now attempt to enrol a much wider audience than was appealed to in the seventeenth century because the funding and governance of science is more complex and is perceived to require public assent. The legacy of Boyle's performative elaboration of the experimental method can be traced through a complex, temporally extended, network of citation practices and literary and visual technologies in the contemporary period that include officially authorised scientific journals in print and now online. Other traces of this legacy can be identified in specific scientific discourses and visual rhetoric that circulate in a wide range of genres and forms.

Furthermore, scientists in general, but particularly life scientists, are increasingly self-conscious about the importance of the media and mediating practices including corporate public relations. For example, it has been widely reported that the team who cloned Dolly held back on reporting on their breakthrough in *Nature* so that they could protect their patent on the procedure. In addition, *Nature* then attempted to manage the international dissemination of this news through issuing a press release about the forthcoming paper with an embargo instructing journalists *not* to report on the story before a certain date. It is also well known that this embargo was, in effect, broken by Robert McKie, science editor of the *Observer* who learned of the story from a TV producer with whom the team at Roslin Institute had co-operated (Kolata 1998: 32; Wilmut and Highfield 2006: 118). These unexpected developments threw into temporary disarray the press relations strategy which had been agreed by the team and their commercial partners (Kolata 1998; Wilmut et al. 2000a; Holliman 2004; Wilmut and Highfield 2006). As this episode illustrates, the imperative to communicate with 'the public' makes high demands on scientists in dealing with the media, particularly in the context of intense competition for resources.

Credibility struggles in cloning and stem-cell research

We contend that the domain of cloning and stem-cell research is being produced in credibility struggles in a public theatre of contestation about gender, nation and science. Such struggles have been a persistent feature of the field in recent years as regulatory frameworks change and experimental results are reported. Nerlich and Clarke make a similar observation with specific reference to the claims made by Severino Antinori and Panos Zavos in March 2001 in what they call: 'a big media event which led to a global debate about bans on human, therapeutic and stem-cell cloning' (Nerlich and Clarke 2003: 45). Nonetheless, given that, to date, neither therapeutic cloning nor human reproductive cloning have fulfilled the claims made for them, this contestation is likely to continue, even if, or, perhaps, *because* the terms of the debate are shifting. As Gieryn argues:

> The selective attribution of this or that characteristic to science cannot be explained by what science 'really' is at the bench or in a journal, but only by the pragmatic utility of any given borders and territories for the protection or expansion or denial of scientific authority over the facts.
>
> (Gieryn 1999: xi–xii)

Credibility struggles about cloning are about such 'authority over facts', which is never a matter of simply reporting experimental successes. (Indeed, no report of experimental success is ever simple.) In the period since the announcement of the birth of Dolly, 'the pragmatic utility' of the borders of genomic science have been rendered explicit. In the name of maintaining legitimate scientific authority, the borders of genomic scientific territory have been erected, stormed and rebuilt. Exemplary scientists have been prime actors in this theatre of contestation and, in effect, they have come to embody these processes.

Second-order virtuality and the figure of the scientist

In his study of the cold fusion farrago, Gieryn reflects:

> The media here became the late twentieth-century equivalent of what Steven Shapin has described as the 'gentleman witnesses' so vital for attesting knowledge claims three centuries earlier. Originally, a gentleman scientist might visit a colleague's lab to examine firsthand his experimental apparatus. This tradition soon gave way to a more removed peer review, as scientists learned about others' experiments through their representations in what would become scientific journals. At first, print accounts contained every minute detail of each instrument or procedure, though with time the scientific paper became increasingly elliptical as more and more parts of the experiment became routine, not requiring elaboration. In a sense, involvement of the mass media in scientific discovery extends that trend toward enlarged ellipses in reporting experimental details. Journalists at Salt Lake City became privileged witnesses to the physicalness of the experiment, privileged not because of their expertise or trustworthiness but because they were needed as conduits for a 'second-order virtuality'.
>
> (Gieryn 1999: 200–1)

With regard to genomics in general, and cloning in particular, we see such 'second-order virtuality' as crucial. It has been assumed that the integrity of the claims to knowledge made by genomic life scientists must be assessed with reference to their embodied performances of the norms of science. Such performances of their expert status correspond to those traced in Shapin and Schaffer's delineation of the gentleman witness, or in Donna Haraway's critical re-appropriation of this figure – the modest witness. In recent developments within human genomic science (and cloning, in particular) the media, supported in many cases by scientists and legislators, have authorised that expert status by distinguishing between reputable and disreputable sources. The claims to knowledge of a reputable source do not appear to require the physical witnessing that would be demanded of a disreputable source.

Virtual witnessing in late twentieth and early twenty-first century science is supposed to be achieved through publication in scientific journals, with second-order virtuality being realised through the further dissemination that results from the interaction between scientific journals and general news media. However, the situation is more complex than this overview suggests. For example, some scientists and scientific institutions endeavour to supplement or to replace this mediating chain by communicating directly with news media. Sometimes such direct communication garners little or no attention since it is regarded as simply part of scientific culture. At other points, this practice becomes a matter for the scrutiny of both scientific communities and the wider public, as it has been associated with problematic scientific practices but, most especially, with problematic scientists. The case of Hwang is an interesting and revealing example of this. His direct address to news media was initially framed as legitimate in the context of global astonishment and excitement about his scientific breakthroughs. However, when Hwang's breakthroughs were shown to be fraudulent, the many media accounts linked his crimes against 'Science' to what were construed as his inappropriate media operations. We will elaborate on this further in the section 'Fallen heroes' (pages 84–9).

Why figure 'Science' or 'science' through individual scientists?

Although Roslynn Haynes's historical overview of representations of scientists in Western literature is centrally preoccupied with the portrayal of scientists in fiction, she also draws on other literary forms including biography. She traces the emergence of the figure of the scientist as a credible social figure, noting that:

> It was primarily through the influence of one person, Isaac Newton (1642–1727), that the popular image of the scientist changed from that of either a stupid or a sinister character to that of a highly respected man of genius, representing the highest attainments of reason. At the time of his death and in the years immediately following, Newton was honoured with the highest literary praise ever accorded a scientist.
>
> (Haynes 1994: 50)[2]

As Haynes's commentary indicates, individual characters, whether fictional or drawn from life, are crucial in any form of narrative. They are agents who mobilise plots and they may become the focus for readers' identifications. Hence, they may represent (embody) or obscure a complex set of actor networks or structures. In modern story-telling about science, in both popular histories and in fiction, individual scientists loom large as discoverers of scientific theories and inventors of technological applications and practices. Such stories draw on archetypal plots of discovery and heroism and, as Haynes notes, centuries of stereotyping. However, it is important to remember that the stereotypes that Haynes identifies are not unremittingly negative, since they include such positive figures as 'the heroic adventurer' and 'the idealist'. Moreover, the deployment of stereotypes (even negative ones), may mobilise or enhance the distinction between 'science' and 'Science' (Stern's terms).

Stereotypical representations have been a feature of popular representations of genetic and genomic scientists in the late twentieth and early twenty-first centuries. This is apparent from the characterisation of Francis Crick and James Watson as the 'discoverers' of the double helix in the 1950s through to the portrayal of John Sulston and Craig Venter as 'racing' to find the map of the human genome in the 1990s. Turney cites Mulkay's detailed analyses of the press coverage of the debate on the UK Human Fertilisation and Embryology

Act (1990) to note that, although invocations of Frankenstein in newspaper reporting formed a very small minority: 'They reinforced the impressions some scientists have that Frankensteinian imagery dominates public discussion of their work.' Despite such critiques, these tropes continue to be drawn on in the framing of news of this field (Mulkay 1997: 116, 130; Turney 1998: 216). Since the birth of Dolly, such tropes have predominantly emerged in the representations of reproductive (rather than therapeutic) cloning. Hence, their mobilisation in the wake of claims by individual scientists, such as those surrounding Dr Zavos's putative clone implantation in 2004, serve to demarcate therapeutic from reproductive cloning. It thus served to identify the former (therapeutic cloning) as *neither* mad nor bad (Haran 2007).

In popular film, the figure of the individual scientist often represents or embodies science. This figuration often references literary precursors such as Prometheus, Frankenstein or Faustus (Haynes 1994; Turney 1998; van Dijck 1998; Kember 2003; Weingart et al. 2003). Despite the diversity of representations within recent films, many contemporary cloning or genetic engineering narratives portray scientists as mavericks, devils or misguided men who have lost their way. Although negative images of scientists abound in modern popular Western film (Haynes 1994; Frayling 2005), genomics has been particularly negatively portrayed. Hence, suspect, evil or mad genomic scientists have been the norm in popular film of the last decade.

The framework of much 'public understanding of science' (PUS) research also highlights the individual scientist as the pivotal figure in science communication and engagement. The assumption which informs many investigations undertaken under this rubric is that ethical and acceptable science is mainly a matter of individual scientists behaving and communicating well. Instrumental or deficit models of PUS highlight the need for individual scientists to convey their knowledge more effectively and more accessibly. Science in Society models place great store in public consultation processes through which scientific research agendas may be reviewed and, to some extent, shaped (Gregory and Miller 1998). So, this model also invests heavily in the promise of individual scientists communicating with 'lay' groups. Thus, individual scientists are generally viewed as carrying responsibility for scientific research and, indeed, as embodying science. Many versions of PUS posit that science is a process mobilised by agential scientists who 'cause' it to happen. In turn, the 'effects' of science are thereby regarded as outcomes which other individuals are free to adopt or refuse, in a neo-liberal model of agential choice.

Hence, it is not surprising to find that individual characters – scientists – are generally portrayed as the source and cause of science in recent film, television drama and docudrama, and television and press news coverage. This means that, on the one hand, science is often represented as de-contextualised and understood as the product of the activities of individual geniuses. On the other hand, fears and fantasies about science can be exorcised through fairly simple plot resolutions in which the individual scientist is disempowered, stopped, or experiences some personal transformation. The practice of 'Science' (in Stern's sense) is thus represented in film as self-regulating and primarily as a matter of personal ethics. The institutions of science and the applications of technology are represented as transparent and distinct from other spheres of life in such figuration of science, through scientists, who appear as a set of individuated agential actors who 'cause it'. In news stories, the range of figures includes larger-than-life scientific icons who carry 'breakthrough' stories and members of an extended peer community who either legitimise or disavow scientific claims. So, in this chapter we will discuss the processes by which Zavos and Hwang ('real life' scientists) were cast as 'mad' and/or 'bad' and accordingly, caught out – not by the action heroes of *The 6th Day* or *The Island* – but by the literary and social technologies that comprise 'Science'.

As already suggested, the figure of the modest witness, constituted by Boyle and interrogated by Shapin and Schaffer, Potter, and Haraway is not the only iconic rendering of the scientist. The scientist has been variously figured as a maverick, madman, god or hero and there is now a significant body of scholarship documenting and analysing these typologies (Haynes 1994; Turney 1998; Frayling 2005). We have chosen to focus on mavericks, madmen and fallen heroes in our analysis because these categories have particular resonance in the developments in cloning science during the last decade. Returning to the iconic figure of Victor Frankenstein, protagonist in the text which is seen as providing the 'governing myth of modern biology' (Turney 1998: 3), we would note that Shelley's narrative deploys each of these stereotypes. Likewise, each of these figures is in evidence in factual and fictional texts on cloning which have appeared since 1996.

The maverick scientist

The maverick scientist is a figure crucial to the management of the boundaries of science and, as such, is a figure of great ambivalence. Positive and negative versions of the figure are produced in a range of genres, but our primary focus in this section will be on the constitution of 'real' contemporary figures as mavericks. Our shift from adjective to noun is deliberate, because the very process of identifying and labelling particular figures as mavericks itself becomes part of the process by which they are excluded from the realm of science. As we shall indicate below, in the field of cloning, where the work of clinicians and of 'pure' scientists is materially contiguous, the designation of 'science' and 'scientist' is particularly flexible.

Since the Raelians failed to produce any evidence of the birth of Baby Eve in early 2003, Zavos is one of the leading 'mavericks' in the story of human cloning.[3] In January 2004, he courted censure from the UK scientific and media communities when he held a press conference at which he claimed to have implanted a freshly cloned human embryo into the womb of an infertile woman. Across the broadsheet, mid-market and tabloid press in the UK, as well as on television news bulletins, Zavos was variously characterised as a 'US fertility maverick' (Revill 2004), 'the clone race maverick' (Woods and Leake

2004), 'Dr Frankenstein' (Thurlbeck 2004) and a 'cowboy cloner' (Cookson 2004). His credentials and interests were framed in such a way as to position him (and his endeavours) as the bad other to the implied figure of a white, British, modest witness. References to his ego, his desire to be first in the clone race, and his financial preying on the desperation of infertile couples undermined his expertise, which was both acknowledged and rendered suspect simultaneously (Laurance 2004; Woods and Leake 2004). This was effected through comments about his disregard for the standards of scientific evidence, his evasion of the culture of publication in appropriate peer-reviewed journals, his dubious ethics, and the potential physical harm that could be experienced by his patients/clients and any putative foetuses or children conceived through cell nuclear replacement cloning (Revill 2004). There appeared to be an enormous discursive over-determination in what became, in effect, the staged expulsion of Zavos from the scientific community. For example, Alex Thomson, presenter of the UK *Channel 4 News*, conducted a live interview with Zavos, on 18 January, 2004. During the interview, he persistently differentiated Zavos from 'proper' scientists in the phrasing of his remarks, which left little opportunity for Zavos to respond and destabilised his credibility as a scientist. Through these diverse modes of media representations, Zavos's maverick status was established and he was clearly located as outside the community of reputable scientists.

Repeated references to the USA, where Zavos was based, and to Cyprus, where he was born, marked him as beyond the UK pale. His reticence to disclose the geographical location of the procedure he claimed to have carried out, which could be interpreted as flaunting the legal strictures on human reproductive cloning in particular nation-states, underscored his outsider status. In these reports, Zavos was 'not Science' and 'Science' was not Zavos. Moreover, human reproductive cloning was constructed as unethical, unsafe and probably (but not certainly) impossible, at least for Zavos and possibly for some years to come.

Jo Revill's account of the press conference and its reception in the *Observer* was one of the most derisive in its framing of Zavos's claims and marshalling of reputable sources to refute them. The account opened with the sentence: 'A US fertility specialist flew into Britain yesterday to announce that he had transferred the first cloned embryo into a woman – but he refused to give a shred of evidence to back up his astonishing boast.' This statement succinctly encapsulates the key elements of the UK media treatment of this story, referring to Zavos as a geographical outsider and a braggart without the proof to substantiate his claims. Revill suggested that the press conference had 'descended into farce when he criticised the highly respected medical journals *Nature* and *Science*, saying he wouldn't want his work to be reviewed or published in them because they did not have enough experts to deal with it' (Revill 2004). As this news event took place before the Hwang scandal, UK journalists were able to take for granted the status hierarchy of scientific journals, and frame Zavos's disrespect of this authority as evidence that he did not abide by the rules of the scientific community. In an interview we conducted with Zavos in March 2007, he suggested two reasons for publishing in journals such as *Fertility and Sterility* and *Archives of Andrology*.[4] First, he claimed that journals such as *Nature* sometimes make decisions about publication based on ethical judgements rather than the quality of the work, and that his papers are disqualified on that basis. Second, he claimed that when he submits articles to journals such as *Fertility and Sterility* and *Archives of Andrology* they are reviewed by appropriate experts. Referring to the January 2004 press conference he said: 'And I said that to the British, and they were laughing at the press conference. I said "How dare you. I like to see what my peers are when they review my paper"' (Interview with authors). In this way he at once responds to and attempts to contest judgements about his expertise and credibility.

Revill also used quotations from multiple UK expert sources to discredit Zavos, including Robin Lovell Badge, 'professor of genetics at the National Institute for Medical Research', Wolff Reik, 'cloning expert at the Babraham Institute in Cambridge' and Professor Alison Murdoch. Murdoch is identified as 'chair of the British Fertility Society' in this account, as the press conference took place prior to the media coverage of her involvement with therapeutic cloning at the Newcastle Centre for Life. The 'chorus of voices' used to discredit Zavos demonstrates the effectiveness of the co-ordinating efforts of the Science Media Centre which had marshalled a rapid response to the news that he was planning to hold a press conference in London.

By staking claims and constituting himself as a pioneer in the field of human reproductive cloning, including displaying mocked-up photographs of himself in a spacesuit walking on the moon at the before-mentioned press conference, Zavos's self-presentation was as a positive maverick, a ground-breaking pioneer. Nevertheless, the experts who commented on his claims, and many of the journalistic commentaries, construed him as mad, bad and dangerous.

The image of the mad/bad scientist is also one which looms large in the cinematic genealogy traced by Christopher Frayling in *Mad, Bad and Dangerous?* Frayling contrasts this figure with the heroic and saintly 'real-life scientists' who were the mythologized protagonists of a string of biopictures in the 1930s and 1940s (Frayling 2005). Frayling launches his study of the portrayal of scientists in Western film, by referring to Roslynn Haynes's (1994) periodised typology of images of scientists that she argues have become 'deeply embedded in the culture [of Western Literature]' (Frayling 2005: 35). Haynes's chart of a sequence of stereotypes of scientists in literature, from the late sixteenth century to the mid-twentieth century, includes: the alchemist, the absent-minded professor, the inhuman rationalist, the heroic adventurer, the helpless scientist and the social idealist. Following Haynes, Frayling suggests that these types 'constitute an image-bank that has been reproduced and redefined in different forms and combinations over the past 500 years' (ibid.: 36).

Although we work with a rather different repertoire of figures from those that Haynes and Frayling present, we do share their interest in the persistent categorisation and stereotyping of scientists in many forms of Western culture.

However, we are concerned, not so much with identifying instances of stereotyping, but rather with tracking and analysing the operations of different literary and visual technologies around these figures and with the relationship between the production and consumption of such imagery. More precisely, we are interested in looking more closely at these mechanisms and relationships in the recent making of human cloning technoscience.

From this vantage point, the media coverage of Panos Zavos constitutes an interesting case study. Zavos was constructed as a negative maverick by reputable scientists and policy makers who were quoted in press and television coverage in the wake of his 2004 announcement. He was denounced for his claims that he was using cell nuclear transfer technology in order to pursue reproductive cloning. However, Zavos attempted to construct himself as a pioneer, in the manner of Steptoe and Edwards who were responsible for the birth of the first IVF child Louise Brown. Indeed, Zavos insisted that his aim, in making his announcement in the UK where this pioneering IVF work was undertaken, was to encourage the appropriate regulation and governance of reproductive cloning. He contended that banning this application of cell nuclear transfer technology would be ineffective and that what was required instead was to ensure that it would be carried out with due oversight. In making such claims, Zavos mobilised the rhetoric of future promise (see Chapter 3) which has come to be a key feature of discourses pertaining to therapeutic cloning in recent years. In framing reproductive cloning as an appropriate treatment for certain types of infertility, Zavos effectively aligned it with therapeutic cloning and thereby challenged the investment in a sharp division between these two forms of cloning.

Although Zavos embraces his maverick status, he is very critical of the way that the UK media report on his work. His criticisms are an interesting mirror image of the critiques levelled at him. He says that the press in England[5] 'do not report the news, [they] just make the news' and is sceptical about the degree to which they report facts rather than write news stories to a pre-set agenda. Frayling suggests that 'science news' is 'nearly always about constructing stories by playing to public worries and setting "agendas of anxiety"' (Frayling 2005: 39). Frayling oscillates intriguingly between evaluating media portrayals of science and scientists against an assumed ideal of rational and balanced communication of 'the facts' and acknowledging that there are a range of competing interests involved in the making of science that play out in the media, as well as in other settings. Hence, he offers the following appraisal:

> What has come into sharper focus, over the years, is the poor regard of scientists for the public's ability to understand scientific issues, and for the media's role in presenting or reporting science . . . scientists' attitudes towards the public, the media, and their own image have themselves become barriers to the communication with the public they themselves seek.
>
> (ibid.: 45–6)

Alok Jha, a science correspondent for the *Guardian*, has echoed Frayling's sentiment and extended the critical musings on how scientists should deal with the public and the media. He did this in response to the publication of a letter from fourteen UK scientists – a letter they wrote to protest about the news coverage of Zavos. In this letter the scientists called upon journalists to cease reporting dubious claims about human reproductive cloning (Cookson 2004; Henderson and Hawkes 2004; Science Media Centre 2004):

> But perhaps these outbursts from the science community, usually best noted for its objective calm, slightly miss the point. If the stories – correct, exaggerated or otherwise – have already been published, and the public is asking about the possibility of reproductive cloning, does that not define the very beginnings of a debate? So is it not better to bring the full weight of the scientific evidence to bear and address public concern head on? And why should scientists demand the controls of any debate that has an element of science attached to it?
>
> (Jha 2004)

The scientists referred to here, whose letter had prompted Jha's comments, had argued that Zavos, like Severino Antinori and the Raelians, has been given too much media attention and that: 'the disproportionate coverage given to these stories convey(s) the impression that fertility scientists in general are engaged in the race to clone the first human.' However, this comment seems somewhat disingenuous or, at the very least, naive, given that Zavos's claims, as we noted above, were framed almost universally by UK journalists as maverick, mendacious and unethical. This aside, Jha's sharp reaction indicates that he is much less concerned than the signatories to this newspaper letter about the risk to public confidence in science posed through coverage of Zavos and other scientists who have made claims in this technoscientific field.

Simon Locke observes succinctly and acutely that: 'Science in general, like any professional practice in a modernised society, needs to maintain public support; traditionally, this support has been sought in part though appeals to the public nature of scientific knowledge' (Locke 1999: 78). However, as Frayling, Turney and many others have noted, news of the human biosciences seems to evoke particularly strong reactions and concerns because they appear to get to the root of what it is to be human. Added to this is the context in which therapeutic cloning has been secured as a possibility in the UK because of the efforts of scientific interest groups to legalise research on human embryos (see Chapter 2). Hence, there are many reasons why 'the scientific community' at that point might have wished to displace or to remove Zavos as the focus of media discussions about cloning (Mulkay 1997; Kitzinger and Williams 2005). In addition to suggesting that he misrepresented reproductive cloning as an endeavour in which mainstream fertility scientists were engaged, the scientists who signed this letter of protest argue that:

> Dr Zavos came here [to the UK] exclusively to get publicity. As with previous similar media announcements, he was richly rewarded, leading almost every news bulletin on television and radio throughout the day. He and others are thus allowed huge air time to prey on the desperate hopes of infertile couples for whom current infertility treatments do not work and on the fears of those who feel that science is out of control.

Zavos himself acknowledges that he holds press conferences in the UK because of the global reach of the UK media: 'I use the British press to make my case. And, of course, as you realise, you do pay a price for that. And I'm willing to pay that price' (interview with authors).

The price, as we have suggested, is that he is represented as operating outside the legitimate scientific community and as, therefore, incapable of making scientific truth claims. The construction of Zavos as a maverick was reiterated and endorsed across the UK press and television news. We see this construction as operating in the interests of those fertility scientists who have formed alliances with researchers working on therapeutic cloning and stem-cell research. Of course, the two groups are not necessarily distinct. Ian Wilmut is a notable example of a fertility scientist – albeit one who worked with non-human animals – who is now pursuing therapeutic cloning. The newspapers assembled sources as diverse as stem-cell scientists, members of parliament and anti-abortion groups to castigate Zavos (see previous references). On Channel 4, the presenter persistently framed him as outside the community of reputable scientists. In this way, Zavos, the maverick, for the time being at least, secured the legitimacy of therapeutic cloning which was understood as a desirable practice, in contrast with his own practice (of reproductive cloning).

Madmen and mad scientists

The creation or manipulation of 'life itself' has frequently been represented as the province of the 'mad scientist'. This formulation applies to some claims by 'real' scientists – most notably claims by Brigitte Boisselier of the Raelians (the religious sect that believes that all human life was created by extra-terrestrial scientists). However, it is most evident in films. The mad scientist, driven to create or manipulate life appears in many popular films loosely based on Mary Shelley's *Frankenstein* through to the various takes on Robert Louis Stephenson's *Dr Jekyll and Mr Hyde* and H. G. Well's *The Island of Dr Moreau*. So it is hardly surprising that recent cloning narratives incorporate the figure of the mad scientist. This is the case in films such as *The 6th Day* and *Godsend*, although in these films the scientist who provides the technical expertise that makes the cloning possible is sometimes absolved of madness through its attribution to those who seek to profit from reproductive cloning. This attribution is important, because it suggests that the idealised institution of 'Science' (Stern 2004) is not threatening, but rather that it becomes problematic when it is appropriated by unscrupulous individuals. Madness appears in these films as a sort of socio-pathology – a pathological disregard of communal mores. In this sense, then, the madmen of these films are mavericks who do not even profess the desire to belong to the same moral community as their fellow professionals or citizens.

In *The 6th Day*, the key scientist is actually a well-meaning genius, and the madman is Michael Drucker, the billionaire owner of the corporation which is illegally cloning humans. Science is both spatially and economically contained by the Drucker character, who is the embodiment of the ruthless industrial entrepreneur. He buys in his scientists and sets them up in high-technology laboratories, hidden (very literally) deep within the commercial buildings that he owns. Thus, science appears as contained and hidden because of the privatising structure of commerce, figured through the megalomaniac Drucker. Scientists and scientific activities are exclusively located in commercial laboratory spaces. Hence, in *The 6th Day* cloning is portrayed as both out of the control of the state or its people and as serving only private interests. It is the conjuring of the dislocation of scientists – since they are removed from authorised and regulated research locations – and of their relocation – in a disreputable commercial setting – that underscores the liberal Christian conservative moral perspective of *The 6th Day.*

The central scientist in *The 6th Day* is Dr Griffin Weir (played by Robert Duvall), a middle-aged, white, male character, who wears a white lab coat to convey his scientific status. He is portrayed as a well-meaning genius, who is, nevertheless, rather blinkered and vulnerable to corruption. Thus, he is unaware of the extent of Michael Drucker's ruthlessness and is preoccupied with his wife's terminal illness and with the prospect of Drucker facilitating the cloning of Weir and his wife. Weir's wife has already died and been cloned once and it is clear that he intends to clone her when she dies again. Weir is eventually redeemed through two moments. The first involves him helping the cloned character – Adam Gibson (played by Arnold Schwarzenegger) – to gain the scientific knowledge he requires to understand what has happened to him and thereby to undermine Michael Drucker's illegal cloning operation. The second is when Weir decides, out of respect for his wife's wish to die, not to clone her again.

In this film's version of science, experimental research is converted into clinical application through the activities of one significant actor and the application of sufficient funding. Through the shots of Weir and his high-technology laboratory, viewers are provided with some glimpses of scientific processes, but these are outside of any institutional setting. Moreover, the narrative establishes that, once the cloning technology is set up, it can be easily executed by non-scientists. As a research scientist Weir has no comprehension of the social implications of his work or links with any larger social networks of scientists or civic actors, and he is thus rendered powerless when he is made aware of his own corruption. One scene in the film depicts a press conference scene. During the press conference, Drucker asserts that Weir is involved in medical research 'not politics', thereby reproducing the conventional cliché that science stands outside of politics. Whilst diegetically Weir eventually comes to recognise his own responsibility in the development of the technology, over which he no longer has any control, he is also positioned as unable to do anything about the exploitation of the techniques he has developed. The film offers a representation of the individual scientist–innovator as a naive and helpless player in the face of capitalist co-option of science, through the development of privatised bioindustries. *The 6th Day*'s Dr Weir is close to the stereotypical scientist Haynes characterises as 'essentially honest and well-meaning but unable to counter the machinations of power politics' (Haynes 1994: 281). What is striking about the deployment of this stereotype in *The 6th Day* is the identification of power politics with corporate capitalism rather than the state, which, in its identification with military applications of science in the wake of the deployment of the atom bomb, provided the monolithic power against which scientists were rendered powerless.

In contrast with *The 6th Day*, *Godsend* revolves around a mad scientist figure, Dr Richard Wells who also becomes the embodiment of supernatural evil as the film narrative shifts from science fiction to the horror genre. In this film the white, male scientist, Dr Wells (played by Robert De Niro) occupies a bright, white and shiny medical environment as a white-coated clinician. However, it is established that he is already isolated from his colleagues in this setting because he uses the clinic to perform an illegal cloning operation, whilst pretending that this is a 'standard' IVF procedure. In stark visual contrast with the shots of his laboratory are those of his home. There he is shown in a dark, dusty room sitting at a huge antique desk framed with purple curtains, with an open fire burning continuously. Robert De Niro's previous on-screen role as Louis Cyphre in *Angel Heart* (1987) is intertextually echoed here. This intertextuality takes on a further dimension in the film merchandising in which the cloning of the child in the film is described as a 'Faustian bargain'.

The professional and the personal are blurred in the film as it becomes clear that Dr Wells has a motive for providing reproductive cloning services that is in excess of the professional ambitions attributed to 'real life' cloning mavericks such as Panos Zavos. This motivation provides the pivot which transforms the film from a science thriller to a schlock horror movie and we will discuss the implications of such genre shifting further in Chapter 6. However, before the film narrative takes this generic turn, Dr Wells appears both as an individual agent and a conflation of the researcher and clinician. This conflation channels and erases the organisational structures of science (in a pattern that is similar to that noted in *The 6th Day*) in that the figure of the scientist stands in for the heterogeneous groups involved in the research processes in actual scientific institutions.

This merging of researcher and clinician is explored further in the film and eventually a moral distinction is made between clinician doctors and research scientists. Dr Wells takes on a variety of medical roles, initially unquestioned by the parents of the cloned boy. As the story develops, Paul (the child's father) becomes increasingly disturbed by, and resistant to, Dr Wells's absolute control over the body of the child, which does not always seem to be in his patient's best interests. The first diegetic doubt about Dr Wells's skills is raised when the child (Adam) wakes up screaming in the night and the father wants to contact the hospital, while the mother wants to contact Dr Wells. This moment triggers tension between the two parents: the mother, Jessie, continues to place her faith in Dr Wells, however, the father, Paul, becomes increasingly anxious to find external support. It soon becomes clear that because of Adam's illegal status as clone, Dr Wells is the only medical practitioner to whom they can turn. In fact, Dr Wells threatens the father with the loss of his son if the doctor's authority over the son's body is undermined. The 'illegality' of cloning, and their willing (initially, at least) participation in this illegal experiment, looms over the parents and thus keeps them from seeking external support and thus imprisons them within Dr Wells's remit.

The explicit reassignment of Wells from the category of clinician to that of research scientist occurs when Paul eventually investigates Dr Wells's dubious past and interviews a nanny who used to work for him. She, a young urban black woman (the only black character in the two films considered in this chapter), relates a story involving Dr Wells's now dead son. She describes Dr Wells: 'he was a baby doctor but he didn't do much doctoring, if you ask me he was a research kind of scientist'. 'Research . . . scientist' is proclaimed in highly disapproving tones, as a clear moral delineation is struck between clinical practice (as curative) and research (as experimentation). Paul articulates this attitude further in a sequence towards the end of the film in which he confronts Dr Wells and accuses him of being an experimental scientist, immorally playing with people's lives, rather than a doctor, who saves lives. Dr Wells is a power-hungry, flawed genius in this narrative. The picture of this character alternates somewhat as he is sometimes presented as a humane genius with an amoral vision and, at other times, he seems to be a corrupt megalomaniac with an immoral secret agenda.

The filmmakers' designation of Dr Weir's 'Godsend Institute' makes explicit reference to his inappropriate attempts to operate as a god through the manipulation of reproduction. In this filmic depiction the character of the doctor/scientist is aligned with the mythical figure Mephistopheles,[6] rather than his dupe, Faust (see discussion of the Faust legend to follow).

Fallen heroes

The earliest of the recurrent stereotypes of fictional scientists that Roslynn Haynes introduces is the alchemist, the best known of whom she suggests is Dr Faustus, whose desire to overstep the god-given limits of human knowledge led him to make a pact with the devil. According to Haynes, Faust, like alchemists more generally, was 'a figure of both fascination and dread, providing an awful example of the moral dangers of intellectual aspirations and pride' (Haynes 1994: 19). Haynes also points out that different aspects of the Faust myth are mobilised in different versions of the tale. In some re-workings of the account, then, Faust takes on the 'moral status of a tragic hero' or his demise is figured as the 'tragic wasted potential of a gifted man' (ibid.: 18).

The recent rise to fame and the subsequent descent into infamy of the South Korean scientist Professor Hwang Woo-Suk could be interpreted as a twenty-first century Faustian tragedy in which a hero is brought down by his own hubris. In February 2004 Hwang and his colleagues published a groundbreaking paper in *Science* announcing that they had cloned thirty human embryos and harvested stem cells from one of them. This was followed, in May 2005, by a further announcement in *Science* that they had established eleven stem-cell lines derived from the skin cells of individual patients (Hwang et al. 2005). Both papers were peer-reviewed and published/publicised in accordance with mainstream scientific protocol. However, at the end of 2005 both were exposed as based on deliberate fabrication.

The way in which Hwang's fall from grace was presented in the UK press echoed Faustian tragedy as the following commentary illustrates:

> Hwang's fall from grace is a nightmare end to what had seemed a fairytale rags to riches story. He worked his way up from rural poverty to become one of the most famous scientists on the planet. In his own country he achieved god-like status with many of his compatriots inspired by his rise from humble beginnings.
>
> (Randerson 2005)

In his early exposure in Western media, Hwang was publicly embraced by other scientists as a *bona fide*, indeed, leading scientist. This embrace took a fairly literal form in press photographs of him posing with his US collaborator Gerald Schatten (Cookson 2005). The research which would eventually be condemned as fraudulent was originally praised by other scientists who, in their commentaries about his research, worked hard to legitimise his research in the face of a potentially suspicious public gaze. In the *Independent:* 'Rudolph Jaenisch, a world authority on cloning and professor of biology at the Whitehead Institute in Massachusetts, said the Korean study was the first time anyone had shown unequivocally that it was possible to clone human embryos using the Dolly technique of "cell nuclear transfer"'(Connor and Arthur 2004). In the *Daily Mail*: 'Roger Pedersen, professor of regenerative medicine, University of Cambridge, said: "The present work has substantially advanced the cause of generating transplantable tissues that exactly match the patient's own immune system"'(Utton 2004). Suzi Leather, then head of the UK's Human Fertilisation and Authority was quoted as saying: 'We can have confidence in their science – these aren't cowboy cloners. I think we can have confidence also in their ethical approach. The Korean research is therapeutic cloning which is very different from reproductive cloning' (Morton 2004). The humble hero origin story was one that had been deployed extremely effectively in the South Korean press and it was also available to readers outside Korea in online, English language, South Korean publications (Tae-gyu 2003, 2005). This trope emerges, albeit with a mildly sceptical framing, in a lengthy piece that appeared in *The Financial Times* in May 2005, in which FT science journalist, Clive Cookson 'explains why scientists in Korea lead the world in human stem-cell work that is putting treatment and cures for devastating diseases and injuries within reach'. Cookson noted that:

> Hwang Woo-Suk, Seoul's scientific superstar, cultivates the media with as much skill as he cultures stem cells. When two FT journalists visit his laboratory, he is happy to dedicate the whole morning to us despite his heavy workload – and he seems seriously disappointed that we cannot stay for lunch.
>
> (Cookson 2005)

However, Cookson had already invoked the modest witness framing of Hwang and his fellow South Korean scientists earlier in the article in offering the following assessment:

> Prof Hwang is not interested in commercialising his research himself. 'I want to remain a pure scientist and I have refused the suggestion of establishing a venture company,' he says. 'We have applied for patents but they will belong to the government and not to our lab.'
>
> Korean stem cell researchers reject vehemently the idea that lower ethical and regulatory standards have allowed them to race ahead. Instead they attribute the country's prowess in stem cell research at least partly to the immense dedication of Korean scientists. 'They are very patient and hard-working people, prepared to sit for many hours working on their stem cells without holidays,' says Prof Moon.
>
> (ibid.)[7]

The South Korean stem-cell breakthroughs triggered general media attention through Hwang's emergence in the mainstream science publishing and publicity route, with articles appearing in high-prestige journals, with accompanying press conferences and press releases. Indeed, various UK journalists drew attention to this conventional pattern, underscoring Hwang's scientific legitimacy. Commentators in *The Times*, for example, contrasted Hwang's announcement with previous spurious claims by 'mavericks' such as Zavos. The South Korean announcement, *The Times* emphasised, was 'the first claim of human cloning that has been verified, peer-reviewed and published in a major journal – the gold-standard for cutting edge research' (Henderson and Hawkes 2004). Nevertheless, when Hwang's ethical misdeeds and scientific fraud were revealed, the transformation of his image in the UK media was rapid. He was repositioned from a reputable and credible scientist to a discredited cloning expert, whose heroic celebrity status could be cited retroactively as evidence that he had been a charlatan all along.

In our discussion of the 'maverick' Panos Zavos, we suggested that reporting secured credibility for the UK regulatory environment, for therapeutic cloning as an experimental practice and for the practices of peer-reviewed science. A different but equally useful process was pursued in response to the Hwang case. When Hwang's team reported its breakthrough, this therapeutic cloning work was established as reputable by contrasting it with Zavos's disreputable reproductive cloning claims and ambitions (see the remark by Suzi Leather above). We have indicated that many UK scientists and legislators have come together in forging a consensus in favour of therapeutic cloning. Hence, they have also been supportive of developments in this arena in other parts of the world, possibly inspired by the ideal of 'Science' as a collective global endeavour. But, because Hwang and his team's research had been conducted outside the UK regulatory framework, virtual witnessing by the UK media and voiced through UK scientific sources was never as secure as it would have been for science conducted within UK borders.

Nevertheless, in South Korea, a local version of the 'modest witness' was repeatedly mobilised. In February 2004, at the time of publication of Hwang and his team's first *Science* paper, *JoonAng Daily* presented Hwang as one of those 'who silently devote themselves to science and to the public' and contrasted him with those 'engaged in deception and cheating with their tongues' (*JoonAng Daily*, 15 February 2004). Following the publication of the second *Science* paper, *JoonAng Ilbo* led with a headline proclaiming that Hwang would refuse to meet with reporters in order to concentrate on his research.

In February 2004, some pre-emptive work was undertaken in UK news coverage to ensure that Korean scientists could not be (mis)understood as maverick simply because of their geographical location. This was effected, in part, through explicit acknowledgement of the instability of the distinction between therapeutic and reproductive cloning. Another key element was the publication of statements by Hwang in which he voiced anxiety about this instability by expressing concern that his work might be taken up by maverick human reproductive cloners for example (Connor and Arthur 2004; Wheldon 2004). In addition to this, as outlined above, in contrast with Zavos, Hwang was presented as the leader of 'a reputable team' that had conducted scientific research in accordance with appropriate protocols and who had published their findings in a suitable peer-reviewed journal. He was thus shown to be a member of an identified community of scientists which was working for the benefit of society. In this story frame, lone mavericks only pursue their own self-interests.

However, the discursive terrain shifted once the research team at the Centre for Life at the University of Newcastle had successfully obtained the first UK licence, in August 2004, to create cell nuclear replacement embryos. Establishing the credibility of the South Korean research in this area was no longer crucial in constructing a progressive narrative about therapeutic cloning. In this rather different context, in order to distinguish between the South Korean work and the new research being conducted in the UK, the UK's status in maintaining the gold standard of regulation and of a rigorous peer community of modest witnesses was foregrounded (Highfield 2004). The *Daily Telegraph* even reported the disputed claim that some of the eggs used by the South Koreans had been donated by a female member of the research team (ibid.). The questions of *how* Hwang had obtained eggs, from *whom*, and how *many*, first attracted suspicion in May 2004 after an investigation by the journal *Nature* (Cyranoski 2004).

However this questioning did not precipitate public scandal until 22 November 2005 when a long-running South Korean investigative television programme, *PD Notebook*, was aired which alleged that Hwang's team had obtained many more eggs than he had admitted, that donors had been paid and not adequately informed of the risks involved in such procedures and, most damning of all, that two of the donors were junior members of Hwang's own team (Chong and Normile 2006). Just before this programme was broadcast, on 11 November 2005, Hwang's USA collaborator (Gerald Schatten) severed associations with him, citing ethical problems. On 24 November 2005, Hwang publicly apologised and resigned from all his official positions. *PD Notebook* planned to air a second segment calling into question the authenticity of the May 2005 *Science* paper, but the Seoul-based television network, Munhwa Broadcasting Corporation (MBC), which airs the programme, suspended its broadcasting, bowing to enormous public pressure and the threat of the withdrawal of advertising revenue. The programme was eventually broadcast on 15 December 2005 as evidence continued to emerge of problems with the 2005 *Science* paper.

The key sequence of events following – or indeed prompted by – the breakthrough stories of therapeutic cloning have now been globally publicised. Through investigative work undertaken by South Korean television researchers, prompted in part by feminist campaigners, as well as by internal whistleblowers and by the investigative activities of young South Korean scientists sharing information and speculation over the internet, Hwang was exposed as a fraud. By the end of March 2006, Hwang's reputation, credibility and career were in tatters (Kim 2006). On 20 March 2006, Hwang was dismissed from Seoul National University for damaging the institution's honour and South Korea's international reputation (Wohn 2006). While scientific fraud and financial mismanagement have been the specific factors in the narrative of this hero's re-presentation as a fraud and a charlatan, the technology of virtual witnessing and the figure of the public scientist have been at the heart of much of the crafting of this story.

Hwang's status as a modest witness, as the leader of a reputable team of scientists who published work in an esteemed, peer-reviewed journal was reiterated in UK press and television coverage of the February 2004 paper. Furthermore, even after unethical egg sourcing and fabricated results had been discovered, his status as a cloning expert was not entirely destroyed. For example, in December 2005, the *Daily Telegraph* and *The Financial Times* called Hwang a 'disgraced cloning expert', or 'once most celebrated cloning expert', alluding to the *change* in his public image, not to doubts about his *status as a scientist* (Highfield 2005a, 2005b, 2005c; Song 2005). It was only the *Guardian* that cast some doubt on his status in January 2006 by putting 'cloning expert' in scare quotes (*Guardian* 2006). In an article in *The Times Higher Educational Supplement*, Stephen Minger was quoted in a commentary that problematised Hwang's media relations but seemed to sustain his scientific credibility:

> Minger, who heads the stem-cell biology laboratory at King's College London, concedes that Hwang's celebrity status might have played a role in his downfall. 'Hwang's group did some good work, but I think it was blown out of all proportion in South Korea. He was a national hero.'
>
> (Fazackerley 2006)

In fact, generally in the UK national media, the focus on Hwang's inappropriate (self-)representation as a scientific hero and his celebrity status were cited to discredit him. For the UK press, Hwang's excessive prominence in, and use of, the media caused as much concern as his scientific wrong-doing.

Gender and modesty

The above discussion has focused on the most high-profile scientists – all of whom are men. This begs the question, what about the female scientist? Despite decades of critique by black and feminist scholars, white, middle-class males predominate in the world of science and they appear in this world as the 'unmarked category'. Although Christopher Frayling's 2003 replication of the celebrated 'Draw-a-Scientist' test devised by D.W. Chambers in 1983 resulted in over 50 per cent of the 7–11-year-old girls sketching female scientists, none of the 7–11-year-old boys drew a female scientist (Frayling 2005: 219). However, it could be argued that the high proportion of female children who imagined female scientists is a more striking phenomenon than their invisibility for male children, bearing in mind the dearth of media representations of women scientists in the late twentieth and early twenty-first centuries.

As we noted in Chapter 2, in recent news coverage on cloning, two female scientists have come to public attention: Dr Brigitte Boisselier, of Clonaid, and Professor Alison Murdoch, of the Newcastle Centre for Life. In the boundary management processes which constitute therapeutic cloning as good science and reproductive cloning as bad science or, even science fiction, Boisselier has become a target for discrediting. Together with Panos Zavos and Severino Antinori, she has been a high profile advocate of human reproductive cloning for many years and like them she has been represented as a maverick. The fact that the organisation that she heads, 'the first human cloning company in the world', is funded by the Raelians (the sect that believes and promotes the idea that human life was created by extraterrestrials), makes her even more of a target for ridicule. However, the more specific representation of her excessive femininity has been a striking discursive vehicle in her discrediting in the UK media. This was particularly apparent in the coverage of her announcement that the first human clone, Eve, had been born on 26 December 2002. Steve Connor of the *Independent* commented that Boisselier had 'the theatrical pose of the Addams Family's Morticia' (Connor 2002). David Adams of *The Times* pointed out that she 'was dressed all in black down to her fishnet stockings, with her hair dyed orange' (Adams 2002), while Julian Borger of the *Guardian* observed that the press conference at which Boisselier announced the birth of baby Eve was made 'all the more surreal by the dramatic orange and white colour scheme of Ms Boisselier's hair' (Borger 2002).

Clonaid's association with the Raelian sect, who claim that extraterrestrial experts in genetics and cell biology had created humans in laboratories, clearly marks Boisselier as a charlatan. In contrast, Professor Alison Murdoch occupies a rather different location: being associated with a reputable, leading-edge bioscientific research unit – The Newcastle Centre for Life. Nevertheless, there has been notable controversy about her status. We have already alluded (in Chapter 2) to Miodrag Stojkovic's reported distancing of himself from what he characterised as 'premature publication' of the fact that Britain's first cloned human embryo had been created at the Newcastle Centre (Templeton 2006).

This public critical distancing of himself from Murdoch has not been his only intervention in framing his former colleague in the public eye. His comments indicating that Alison Murdoch was an inappropriate media spokesperson because she was not directly involved in laboratory work or stem-cell research have also garnered media attention (ibid.). Murdoch, whose primary professional role is as the provider of (in)fertility services, was thereby cast as a clinician (as well as a provider of human eggs), not a scientist. Significantly, however, in the application for a cell nuclear transfer licence to the HFEA, Murdoch was initially named as both the 'Person Responsible for the Centre's Treatment Licence' and the 'Person Responsible for the Centre's Research Licence'. When the licence application was resubmitted following advice from the HFEA 'that it was not appropriate for the same person to be in charge of both the treatment and research activities at any one centre', Murdoch nominated another person at the Centre for Life to be the Person Responsible for the Treatment Licence so that she could apply to be the research licence holder. So the HFEA licence explicitly constructs Murdoch as a scientist, rather than a clinician. Interestingly, the manoeuvring around Murdoch's status contrasts with the positioning of Dr Wells in *Godsend*. In a reversal of the terms of assessment employed in the reported assessment of Murdoch, Wells's experimentation in reproductive cloning was seen as discrediting his status as a clinician, rendering him as a mad or bad scientist.

A feature article in the *Observer* that positioned Stojkovic as, at one and the same time, a scientific hero and a modest witness is striking in its use of visual illustrations. As the reputable male scientist, he is visually foregrounded at the top right of the page, looking contemplative, wearing a white lab coat and rubber gloves. A much smaller headshot of Hwang wearing what look like surgical scrubs is photo-montaged into the shot beneath Stojkovic's (extremely large) microscope. Alison Murdoch's photograph is at the bottom left of the page and is entirely stripped of any visual context. She is wearing a pink low-necked jacket with a small pearl necklace. Her lips are colour matched to the jacket. She is well-coiffed and her head is slightly tilted away from the viewer. Semiotically she bears no connotations of the laboratory or the clinic, and her femininity is of a very respectable, non-threatening variety. Unlike Boisselier whose inappropriate femininity (excessive, or even monstrous) was used to discredit her, Murdoch's much more appropriate femininity is used to code her as subordinate to the scientific genius Stokjovic and as a public relations functionary rather than a public scientist.

Conclusion

In this chapter we have argued that the figure of the modest witness, Haraway's critical re-appropriation of the gentleman witnesses described by Shapin and Schaffer, is the individuated figure who represents ideal and idealised 'Science' (Stern 2004) in contemporary scientific narratives. This individuation is in interesting tension with the emphasis placed on teamwork or community validation in scientists' figuring of the appropriate *modus operandi* for properly conducted science. We have discussed the ways in which mediation was always already implicated in the figuring of the modest witness through the technologies devised by Boyle to produce experimental philosophy, the foundation of modern science. Mediation enabled the extension of identification with the modest witness through the process of virtual witnessing, originally confined to the circulation of specialist publications amongst a select few. Drawing on Gieryn's notion of 'second-order virtuality', we have discussed some of the ways in which the mass media now enable the production of science through the involvement of an extended community of virtual witnesses.

Moving on to explore the figures of the maverick, the madman and the fallen hero, we have explored case-studies of the media production of cloning science and suggested that these figures have been crucial in the identification of appropriate modest witnesses and in the demarcation of the boundaries of good science. We have argued that, in the recent global context, the hegemonic version of good science includes therapeutic cloning. We have also indicated that this inclusion is legitimated and confirmed through the discursive branding (in effect, the expulsion) of human reproductive cloning as bad science. The scientists that have figured in the world of cloning technoscience whom we have studied in this chapter, as we have shown, are overwhelmingly gendered as masculine. Hence, as we have demonstrated, the capacity for positive representations of women scientists in this field has been tightly circumscribed. Thus, even in the new and dynamic world of cloning science, women scientists still seem to constitute what Haraway has called 'an oxymoronic social subject' (Haraway 1989: 281).

The case studies presented in this chapter demonstrate how scientists emerge as powerful agents and representatives or embodiments of 'Science' in the media. This chapter has told stories of celebrity scientists, but it has also considered the discrediting of scientists. Indeed, we have contended that negative representations of particular scientists may play a very positive role in the making of science. In fact, we have shown that the negative figuring of some science and some scientists may provide other scientists with crucial resources for staking claims to reputation, funding and enabling structures of governance.

5 Women's bodies in cloning discourses

Introduction

Our focus in this chapter is on women's bodies and how they have been constituted as the subjects and objects of the discursive and material practices of cloning. We examine how technoscientific cloning is being made through the discursive and material mobilisation of women's bodies. We highlight the historical contingency of the gendering of genomic science by tracing the mediation of human cloning in relation to reproductive technologies as well as genomics. We look particularly closely at developments since 2000 as cloning experiments and licences/consultations around the sourcing of human eggs have become practices in the global technoscience of cloning.

This chapter constitutes an analysis of the ways in which women are emphasised or obscured, in news stories, policy documents, and films, as the source or carriers of eggs and embryos. This involves explicating how this varies within genres and tracing the differences between representations of reproductive and of therapeutic cloning. We explore how particular visual and textual discourses around cloning may destabilise or undermine particular understandings of gendered social relations, scientific enterprise and desirable futures. Our investigation also considers how specific figures and tropes, such as the embryo, travel across different genres and the way meaning is constituted through continuities and intertextualities between film, news and science and policy communication.

The chapter concludes with an examination of the feminist frameworks with which we have engaged. We reflect on our attempts to bring together an analysis of current discourses of human cloning with debates about new and assisted reproduction, the boundary-making processes of science and the global iconography of the life sciences.

How do women figure in discourses of human cloning?

We have already argued (particularly in Chapters 2 and 3) that in the early twenty-first century, human cloning has increasingly been represented as multiply reproductive. On the one hand, it has been identified (usually positively) as a process which provides cures in regenerative medicine and stem-cell research. On the other hand, it has been identified (usually negatively) as a form of new reproductive technology. In the recent constitution and circulation of cloning discourses, women's bodies have frequently acted as boundary figures which are mobilised, conscripted, valorised and elided at different points in contestations around the meaning of cloning and in the negotiations of specific scientific projects. Although women's bodies are our central focus here, issues of national identity, sexuality and capital also emerge in this chapter.

Drawing on the work of Thomas Gieryn (Gieryn 1999), Donna Haraway (Haraway 1997), as well as Sarah Franklin, Celia Lury and Jackie Stacey (Franklin et al. 2000), we develop arguments in this chapter about how women's bodies operate as boundary figures in the discourses of cloning in three distinct and sometimes contradictory ways. They are made invisible and visible in the following ways:

1 They appear as a resource – a source material – which is crucial in constituting science as a national and global project.
2 They appear as the (potential) recipients of cures and reproductive technologies and as carers of the recipients of cures.
3 Their corporeality is erased.

We highlight these gendered regimes of human cloning at a moment when the applications of cloning are neither fully culturally sanctioned nor naturalised.[1] Our concern is to extend understandings of recent patterns in the making of gender and of cloning technoscience. More precisely, we show how, through their appearance and disappearance, women's bodies operate as intelligible figures of reference, demarcating acceptable from unacceptable forms of cloning. In addition, we highlight moments of disruption and rupture in cloning discourses in which women's bodies are pivotal. We also suggest that the naturalisations and occlusions in the recent discourses of cloning, draw on discursive repertoires and political patterns that have been crucial in the normalisation of IVF and embryo research (particularly in the UK). We explain in the conclusion that we anticipate that these may help to normalise and extend experimentation in technoscientific cloning.

Different texts and genres employ cloning terminology in different ways and technical and rhetorical distinctions and features can be crucial to understanding developments in this field. For example, in factual genres, the distinctions between terms such as blastocyst (pre-embryo), embryo, foetus and human subject are all crucial. Moreover, the distinction between identical genetic material and identical physiology is also significant in cloning technoscience, although this is not always easily or clearly represented, and the conflation between the two is particularly dominant in filmic representations of cloning. Furthermore, as we have shown previously, while news genres have emphasised the distinction between reproductive and therapeutic cloning, this distinction has been much less obvious in film.

All these terms and distinctions have either emerged and/or been subject to reconstruction and reformulation during the period of our study. They also carry different kinds of ideological connotations and, as we shall explore in this chapter, it is through these that cloning acquires its meanings. In addition, there is a complex array of discursive legacies which have contributed to recent cloning discourses. A partial list of these could include: the development of IVF since the late 1970s; the embryo debates in the UK in the 1980s (Crowe 1990; Mulkay 1994); the rhetoric of genomics in the 1990s (Nelkin and Lindee 1995; van Dijck 1998; Nerlich et al. 2002; Nerlich and Hellsten 2004); and the reconfigurations of the embryo in the stem-cell debates of the late twentieth and early twenty-first centuries (Nerlich and Clarke 2003; Parry 2003; Kitzinger and Williams 2005; Franklin 2006). These are the sites and discursive terrains in which the core technologies (Throsby 2004; Thompson 2005) and the key discursive registers which inform and constitute human cloning have emerged. Genetic counselling, genetic testing, stem-cell research, therapeutic and reproductive cloning are some of the key, interconnected sites in which women's minds and bodies come under new forms of biotechno-scientific scrutiny and investigation. This chapter provides both an overview of the appearance and disappearance of women's bodies in the proliferating sites of human cloning and a focused analysis of women's bodies as they appear and reappear in specific texts.

Revisiting terminology and legislation

Therapeutic cloning is currently represented as a process for generating embryonic and DNA-matched stem cells through the production of somatic cell nuclear transfer (SCNT) embryos. The meaning of the term 'cloned embryo' has been subject to contestation with new terms such as the 'reprogrammed cell', the 'derived embryo' and 'embryos not formed through fertilisation' entering the discursive terrain. These developments are embedded in established discourses about abortion, embryo research, and IVF as well as stem-cell research. However, as we have noted previously, the coupling of stem cells with therapeutic cloning has added a further dimension to cloning. Thus SCNT is associated with regenerative medicine and tissue transplantation, in addition to its association with reproductive medicine and infertility. We have traced the emergence of the distinction between reproductive cloning and therapeutic cloning in the wake of public consultation about the possibilities of human cloning after the news regarding the birth of Dolly the sheep (see Chapter 2). We have also emphasised the instability of this distinction. Indeed, reproduction and therapy became inextricable from the start of the 'clone age', since the reproductive cloning experiment that was Dolly was used to herald the prospect of the bioeconomic goods of regenerative medicine and drugs (Franklin 2003). The fact that the international policy impact of mammalian cloning was that governments sought to generate or revise their human cloning legislation undercuts the claim that Dolly could *not* be a precursor to human reproductive cloning (Highfield and Wilmut 2006).

The cloning of human embryos using SCNR (somatic cell nuclear replacement) or SCNT has become associated with the elimination of 'fertilisation'. This is because there is no direct use of male gametes and because SCNT becomes the motor for provoking cell division. A woman's egg is enucleated – the nucleus is removed – and the original nucleus is replaced with one transferred from an adult or somatic cell. One of the potential benefits of embryonic stem cells produced in this fashion is that potentially they can be DNA-matched with a donor who provides the somatic cell (the nucleus, not the enucleated egg). This matching could overcome problems of rejection when these manipulated cells are reintroduced into the body. However, the distinction between IVF embryos (fertilised) and SCNT embryos is sometimes highlighted and sometimes conflated in the debate. This distinction has thus become another strategic rhetorical distinction (like therapeutic or spare) that may either be accentuated or elided.[2]

We have explored some aspects of the construction of the distinction between therapeutic and reproductive cloning, the reconstruction of the embryo as therapeutic and the links to stem-cell research in earlier chapters, especially Chapter 2. However, in this chapter we are specifically concerned with the dislocations, the de- and re-contextualisations and the disavowals at work in these developments. The preceding review helps to illustrate the discursive work that is going on around developments in cloning, particularly in relation to women's bodies. We are especially interested in the discursive work that polices the boundary between therapeutic and reproductive cloning and between the technoscientific and the corporeal. We examine the making of these differences and separations, as therapeutic cloning comes under assault from the spectres of reproductive cloning, the labour of the body, the exploitation of women, and scientific fraud.

During the period of our study, debates around how women's eggs are to be secured and employed in the process of enucleation have emerged in different ways in South Korea, the USA and the UK. In 2006 and in early 2007, the USA and the UK were engaged in deliberations about how eggs could be obtained for therapeutic cloning research and about appropriate legislation to ensure the availability of this material for this purpose. In the wake of South Korea's temporary emergence as the world leader in therapeutic cloning, issues around women and eggs were raised at specific key moments which we examine in the following sections.

Although South Korea and the UK have occupied centre stage in the UK media coverage of cloning and human genomic science, developments in the USA have also garnered attention. We have noted that, in contrast with the UK, the US government banned federal funding for embryo research. Nevertheless, it has never legislated to prohibit human reproductive cloning. In addition, there has been coverage of the secession of some states, such as California and Massachusetts, from the ban on federal research funding, and of the pledge of public money at state level to fund human embryonic stem cell (hESC) research. California has been particularly controversial in that regard as arguments over funding, spending and stem-cell research escalated there throughout 2005.[3] California has also emerged as a potential human biotechnology partner for the UK. The UK media gave considerable attention to Tony Blair's visit there in July 2006 and to his meetings with the California-based, biotechnological companies Genentech, Gilead Sciences and Cell Genesys.[4]

There is significant opposition to human biotechnology on religious grounds in the USA and George W. Bush's stance, expounded in his *State of the Union* speech of 2006, has heightened the controversy around cloning and stem-cell research in this context. We suggest that the latter may even have precipitated some backlash in the form of intensified interest and investment in stem-cell research. The UK is usually represented in reporting on this field of biotechnology as a comparatively more secular nation, which is less embroiled in controversy about the regulation of embryo and stem-cell research. President Bush himself is sometimes cited in the UK press as demonstrating how *not* to regulate this research: 'Bush and co. don't care about whole big people. They would rather focus on a few clumps of cells' (Lionel Schriver, *Guardian*, 27 July 2006). Nevertheless, in the mid-term elections of 2006, stem-cell research became a highly visible agenda item in the USA. Since then, the issue has become even more politicised and efforts to increase funding for this techoscientific field are currently underway.

The differences in national regulatory regimes are significant because the bodies and bodily tissues used for this research – eggs and embryos – and the bodily tissues resulting from this research – embryonic stem cells – are trafficked across national borders. Indeed, despite attempts to protect bodies and body parts from the market, international trading in body parts, organs and tissues is now well established and a number of scholars have studied these practices (Andrews and Nelkin 2001; Waldby 2002; Goodwin 2006; Mitchell and Waldby 2006; Nahman 2006; Waldby and Mitchell 2006). Human eggs and embryos used in the production of SCNT embryos and stem-cell lines are also caught up in this trading. So, if cloning and stem-cell technoscience advances as promised, this trafficking in gametes is likely to intensify. The variation in regulatory regimes opens the door for the exploitation of poorer states by richer ones and of poorer individuals by richer ones. This poses particular concerns in relation to the exploitation of women who are the only source of human eggs. So, for example, in relation to IVF, Germany currently imports eggs from Israel and Israel imports them from Romania (Nahman 2006). The UK has also imported eggs and embryos from Romania (for IVF) and Sweden (for cloning), with the regulatory oversight of the HFEA. There are already existing and nascent egg and embryo markets related to IVF (Nahman 2006; Bharadwaj 2007) which are both extending and developing further in relation to therapeutic cloning.

In 2006 the HFEA conducted a public consultation in the UK on the issue of sourcing eggs for biomedical research with a document entitled 'Donating eggs for research: safeguarding donors'. This consultation solicited the views of 'the public and interested parties' (p. 1) to determine the most acceptable forms of regulation for the egg sourcing required for cloning experiments. On the policy agenda were two options: on the one hand, 'altruistic' donation or, on the other, a commercial market. The HFEA consultation appeared to be pushing towards a middle ground in which 'altruistic' donation could also be rewarded, whilst overt commercialisation would be avoided.

In both the USA and UK, at this point, there seemed to be at least a temporary consensus, evidenced in consultation documents and legislation, that an overtly commercial egg market for therapeutic cloning was not desirable. However, the practice of offering cheap IVF in return for egg donation has already been licensed in the UK (July 2006) through the North East England Stem Cell Institute (NESCI).[5] Although the HFEA has said that they may retract this licence, this option was included as the preferred method for sourcing eggs in their consultation documents.[6] In April 2006 California produced a Senate bill (SB 1260) regulating egg sourcing, through which it also tried to avoid creating an explicitly commercialised egg market, and designated women who donate eggs as 'research subjects'. This legislation, like the UK consultation process, invokes the World Medical Association Declaration of Helsinki on research on human subjects (WMA.net).

In South Korea the NGO coalition, WomenLink, has also been lobbying for this issue to be at the centre of national science policy. This lobbying intensified in the wake of the disclosures about Hwang's problematic methods for securing eggs for experimentation: including allegations that he had used 2,000 eggs, that egg 'donors' had been paid and inadequately informed of the risks and that some of these women were junior members of Hwang's own research team. WomenLink held an international conference on 20–21 September 2006 entitled 'Envisioning the Human Rights of Women in the Age of Biotechnology and Science' (WomenLink; Paik 2006).[7]

Science is often strategically represented as universal and thus as based on global collaborations. As Charis Thompson (2005) points out, scientific training and methods are trafficked across geopolitical borders. However, leading-edge science is also often identified with a healthy modern nation. Thus, the practices of sourcing women's body parts for cloning and stem-cell work, as we examine in detail below, require complex discursive labour to make them acceptable and intelligible as simultaneously both global and national projects. This work operates through the figuring of women as both resources (particularly *corporeal* resources) for, and as recipients of, scientific goods. Nevertheless, at some moments, this discursive labour results in the complete erasure of women's bodies. We trace these discursive patterns in more detail in the following sections of this chapter.

Figuring women's bodies

Women's bodies as a resource or source material for science as a national and global project

In this section we examine how women's bodies appear as a resource or source material for science as a national and global project. In the gendered national imaginaries around cloning, women's bodies have been invoked to secure the borders between good national science and science that happens elsewhere. Resilient national imaginaries have been in operation in the UK reporting on Dr Panos Zavos ('maverick cloner'), Professor Hwang ('fallen hero') and Professor Alison Murdoch (the respectable face of well-regulated UK cloning, based in Newcastle).[8] These have worked to ensure that the claims by Zavos and Hwang would shore up UK scientific achievements, whilst simultaneously marking UK science as distinct. In the stories about Zavos and Hwang, women's bodies appeared as corporeal obstacles to the materialisation of scientific human cloning. In stories about Murdoch's work in Newcastle, UK, women bodies appeared as material resources or were erased to support the legitimacy of this work.

Women's bodies were foregrounded in coverage of Zavos in 2004 and their mobilisation both signalled cloning as heterosexual and vilified Zavos as exploiting women. In the UK media's portrayal of the disreputable cloning science of Zavos, and in much recent Hollywood film, the materiality of the female body is used in the portrayal of human reproductive cloning as bad, exploitative, foreign, or fraudulent (or some combination of these attributes). Women's bodies also become the touchstone through which a 'moral high ground' is secured in global IVF and cloning discourses. Thus, IVF and latterly therapeutic cloning is represented as meeting the needs of women or providing them with choices. The purity of science as a global enterprise is also at stake in these representations. Hence, the desire to protect women and, in particular, women's bodies, provides the grounds for prohibiting cloning and for representing it as horrific.

Women's bodies appear as highly visible resources for cloning in order to destabilise the claim to scientific legitimacy of some kinds of cloning, in this case reproductive. At the same time they are also invoked to naturalise reproductive cloning as heterosexual. The figure of the mother (and the father) is used to contain cloning within a normalised version of family relations, invoking narratives about naturalised desires to have children or to replace a lost child. These invocations work as a bid to make the process of cloning more intelligible and to offset anxiety around it. This leaves open the possibility of the recuperation of reproductive human cloning, the horror of which is also undermined by its alignment with IVF. The materiality of the female body thus disrupts narratives of disembodied science, whilst acting as the boundary figure through which science operates.

The UK news coverage of the announcement by Zavos in 2004 that he had cloned an embryo and transferred it seems to indicate that there had been confusion about how to frame this story. The confusion can be traced, in part, to the expectations (outlined in Chapter 1) on the part of both news producers and news consumers that this kind of cloning belonged to fictional genres. The resulting stories conveyed an overload of information and multiple attempts to secure meaning. In this context, the suggestion that reproductive cloning might be a possible 'cure' for infertility did not seem outrageous. In fact, such a suggestion was just one strategy in the many attempts to make meaning about cloning found in this coverage and was used in attempts to describe the process with asides such as: 'the mother is said to suffer from fertility problems' (Figure 5.1).

In the account of reproductive cloning provided in the UK's *Daily Mail* on 19 January 2004, much more attention was given to the way that the procedure is enacted through women than has generally been the case in reports of therapeutic cloning. In this article, questions were raised about the ethics and potential health risks of reproductive cloning. This occurred because reproductive cloning still operates figuratively as outside the borders of the law (in the UK) and of good science. Indeed, this story raised questions about ethics and risks that generally have not been raised in coverage of UK therapeutic cloning announcements. Unusually, anti-abortion campaigners' commentaries (LIFE) were positively framed in terms of the risks to women, and they were represented as on the side of science. This was striking since 'pro-life' campaigners have generally been represented as emotional, religious and oppositional to science. They have often been introduced to provide an alternative perspective to that offered by scientists. However, in this coverage of Zavos's claims, the 'pro-life' campaigner was allied with good science and 'spoke' against bad science. Thus, the reporting on reproductive cloning and the figuring of women's bodies as a resource seemed to open up discursive opportunities to ask questions about risks because it is construed as 'bad' science.

In the article in Figure 5.1, the imaginary of cloning as a heterosexual reproductive technology is evidenced in the image of the male and female figures with their baby. The corporeality of the women's body is emphasised through her nakedness contrasting with the man in the suit. In addition to the heterosexual grammar of the visual images, the text describes the respective donors of the cell and the nucleus for transfer, as 'mother' and 'father'. Despite this form of naturalisation, the iconic image of the foetus is represented as 'alien' in morphology and monstrous through its relative size. It is also identified through the explanatory text as 'twin of his father' (with the naturalisation, the disruption and the uncanny aspects that this implies). The illustration is a vivid, visual representation of attempts to re-naturalise cloning as heterosexual reproduction. The diagram literally contains cloning between a (clothed) man and a (naked) woman. The moment of fertilisation is replaced with 'electrical stimulation'. However the end result, the baby, is superimposed on an image of the couple positioned closer together. At the same time as operating to vilify Zavos (for exploiting women), this focus on heterosexuality

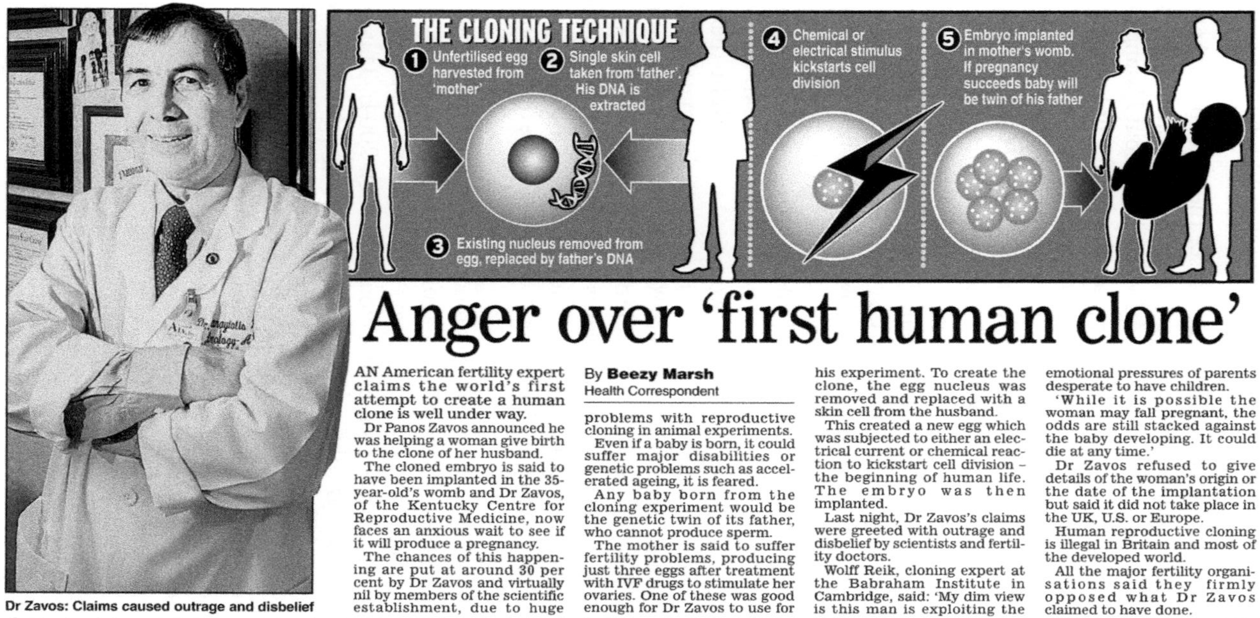

THE CLONING TECHNIQUE

1 Unfertilised egg harvested from 'mother'

2 Single skin cell taken from 'father'. His DNA is extracted

3 Existing nucleus removed from egg, replaced by father's DNA

4 Chemical or electrical stimulus kickstarts cell division

5 Embryo implanted in mother's womb. If pregnancy succeeds baby will be twin of his father

Dr Zavos: Claims caused outrage and disbelief

Anger over 'first human clone'

By **Beezy Marsh**
Health Correspondent

AN American fertility expert claims the world's first attempt to create a human clone is well under way.

Dr Panos Zavos announced he was helping a woman give birth to the clone of her husband.

The cloned embryo is said to have been implanted in the 35-year-old's womb and Dr Zavos, of the Kentucky Centre for Reproductive Medicine, now faces an anxious wait to see if it will produce a pregnancy.

The chances of this happening are put at around 30 per cent by Dr Zavos and virtually nil by members of the scientific establishment, due to huge problems with reproductive cloning in animal experiments.

Even if a baby is born, it could suffer major disabilities or genetic problems such as accelerated ageing, it is feared.

Any baby born from the cloning experiment would be the genetic twin of its father, who cannot produce sperm.

The mother is said to suffer fertility problems, producing just three eggs after treatment with IVF drugs to stimulate her ovaries. One of these was good enough for Dr Zavos to use for his experiment. To create the clone, the egg nucleus was removed and replaced with a skin cell from the husband.

This created a new egg which was subjected to either an electrical current or chemical reaction to kickstart cell division – the beginning of human life. The embryo was then implanted.

Last night, Dr Zavos's claims were greeted with outrage and disbelief by scientists and fertility doctors.

Wolff Reik, cloning expert at the Babraham Institute in Cambridge, said: 'My dim view is this man is exploiting the emotional pressures of parents desperate to have children.

'While it is possible the woman may fall pregnant, the odds are still stacked against the baby developing. It could die at any time.'

Dr Zavos refused to give details of the woman's origin or the date of the implantation but said it did not take place in the UK, U.S. or Europe.

Human reproductive cloning is illegal in Britain and most of the developed world.

All the major fertility organisations said they firmly opposed what Dr Zavos claimed to have done.

Figure 5.1 'Anger over "first human clone"' by Beezy Marsh, Health Correspondent, *Daily Mail*, 19 January 2004, p. 31. Photograph of Dr Panos Zavos by Paul Harris.

and women's bodies also offers a flexibility which keeps open the possible recuperation of reproductive cloning. This is important in news reporting because human reproductive cloning, whether desirable or not, is increasingly anticipated as a technoscientific practice.

The Hwang case, also reported on in 2004 (see the timeline in Appendix I for clarification on the order of these events), mobilised discourse about women in some overlapping ways. In the press and TV coverage about Hwang's experiments with therapeutic cloning, women appeared as a different kind of resource for science, but there were some similar dynamics. In early versions, they were not central and later, in the fraud scandal, they were erased again. Early in the reporting about South Korean human cloning experimental achievements, when the work was constituted as credible science, the fragmentation of the body and of body parts that characterises therapeutic cloning stories was in evidence. The heterosexual coding of cloning also featured in these accounts.

Let us return to a diagram originally discussed in Chapter 2 (Figure 2.1). Our aim here is to revisit it with the focus now on how women are represented. The diagram which accompanies the report on the Hwang breakthrough in the *Independent* (Friday 13 February 2004: 5) shows the egg which is represented as being donated for research. The egg is labelled 'mother'. In the diagram there are at least two distinct dislocated women – at top right there is the 'mother' who provides an unfertilised egg. This raises the question: In what sense can an egg donor be named a mother? We would suggest that the mother exists only at the level of text – albeit bold text – but she's invisible as a body. In the bottom left of the frame the reader is told that the 'Embryo could be implanted into foster mother to produce a child', but the umbilical cord of the illustrated child just fades away. The foster mother – like the egg donor cannot be seen.

The South Korean 'breakthrough' was reported across the broadsheet, mid-market and tabloid press in the UK. In another article covering the Hwang story, this time in the *Daily Mail*, a diagram entitled 'How therapeutic cloning works' was used to illustrate therapeutic cloning (Utton 2004). This explanatory diagram provided an illustration of the donor egg, the adult cell and the resulting blastocyst. In this attempt to report cloning as therapeutic, the language of reproduction was emphasised by references to the 'mother' and the 'embryo' in the article. In the explanatory diagram a silhouette of a naked female body (also used in the image above to explain reproductive cloning) was positioned at the top of the image. The flow of the diagram positioned the naked woman as the source of the egg. This image was dislocated from any context and placed next to an image of the egg and a reference to the embryo was made in the diagram. The naked silhouette and the egg were framed as equivalent through this juxtaposition, making it visually explicit that the body was there as the source of the egg.

To summarise these two articles on the Hwang case; in the *Independent* (Figure 2.1) women's bodies were visually erased but the egg was labelled the 'mother' through textual anchorage. In the *Daily Mail* women's bodies appear as naked silhouettes, equivalent to eggs. These reiterations represent women's bodies as a passive, dislocated source of eggs.

The feminist organisation WomenLink fought against allowing cloning via street picketing, demonstrations, public hearings and press releases. Even before the egg scandal broke, they were particularly concerned about the potential exploitation of poor women: 'Medical technology using embryo cloning technology would inevitably instrumentalise women's bodies' argued Kim Sang-Hee, spokeswoman for the organisation. The media often misses the point that human eggs are part of women's bodies and that: 'the point of the debate is not about the *status of eggs*, but about the *source* of egg' (Interview, summer 2005, conducted and translated by Choon Key Chekar).

In the media coverage in South Korea, women's bodies appeared as a visible disruption to the science of therapeutic cloning through the documentary *PD Notebook* (2005) which exposed evidence about the problems with Hwang's methods for obtaining eggs. However, *PD Notebook* was an exceptional text in the context of the earlier stages of reporting on Hwang in South Korea. Women's bodies appeared more frequently as positive resources for national science with egg donation framed as a patriotic duty (Chekar and Kitzinger, 2007; Jeong 2006). The issue of eggs has remained significant in South Korea, whilst in the UK and USA versions of this story, this issue receded rapidly as issues of scientific fraud took centre stage.

In the long unravelling of the South Korean claims, questions about the source of eggs initially cast doubts on the research (Cyranoski 2004). Although already raised as an issue within South Korea, they were at first only alluded to in the simultaneous announcements by Seoul University and Newcastle in May 2005. However, by November 2005 the USA scientist Professor Gerald Schatten disassociated himself formally from Hwang on the basis of suspect sourcing. Thus, although doubts had been raised about sourcing, it was the admission that his images of stem-cell lines had been fabricated which raised doubt in the scientific community (*Science*, 310, 16 December 2005). Although initially described as a mistake, it was both the fabrication of images and the later evidence that other data had also been falsified which led to the retraction of the scientific article. Hence, it appeared that the fabrication of images and data constituted a greater crime against the science than did issues of consent or exploitation.

News coverage of the Newcastle Centre for Life in the UK and of licences issued by the HFEA, framed women differently again. In this coverage women were framed as protected. This framing was made possible first through the claim that this Centre would not obtain eggs unethically, and second by establishing that it would not allow (women to be exploited by) reproductive cloning. In this discourse of cloning science as a UK project, women's bodies operate to secure both the nation and the science as benign and safe, 'pure' rather than dangerous in relation to cloning (Douglas 1966). However, the claiming of national scientific purity is not confined to the UK. The media outlets of other nations also represent science as pure within their national contexts. While narratives about science as a national project evoke the protection of women, women's bodies simultaneously appear as a resource for science and the nation. In these appearances, as resources, women's bodies are fragmented into oocytes, eggs, embryos, cells, blastocysts, foetuses and wombs. In therapeutic cloning discourses women's bodies are separated into parts or represented as transparent or figuratively, for example as 'mother' cells. Women are figured as lacking in agency and also dislocated.[9] Such representations construct therapeutic cloning as a technoscientific practice which is simultaneously removed from women as subjects, yet dependent on the materiality of the body.

When women's agency or subjectivity has been invoked, this has been in relation to their willingness and 'duty' to donate. This is a construction which has been used by Ian Wilmut and Alison Murdoch and surfaces in the HFEA consultation on egg donation of 2006 which refers to an imagined 'moral obligation to participate in medical research' (HFEA 2006: 10). Hence, women have been represented as subjects when they appear in subject positions which affirm cloning science. So, in the initially positive coverage of the 'breakthroughs' in South Korea, discussed above, women appear as fulfilling their patriotic duty to science as donors.

Alison Murdoch reflected on the relationship between women and science in the following way:

> Alison Murdoch: Yeah, and we say that to our patients; most of our patients who donate eggs and embryos say 'if we haven't . . . if someone hadn't done the research in the first place, they wouldn't be here having a chance to have a baby'.
>
> Interviewer: Right.
>
> AM: And they understand that, and on that basis they agree to give their cells for research, and they are the people who we really should be listening to, because it's their cells.
>
> (interview with authors, 7 November 2005)

In the 2006 HFEA consultation document 'Donating Eggs for Research', women only appear when they are represented as willing donors of eggs. In this document in the section that invites a response, women are only named twice and on both occasions it is to *affirm* egg donation. In the rest of the document, in which there is some opportunity to raise questions about egg donation, 'women' are replaced in the text by more generalised terms such as 'people' and 'donors'. It is only in the final section that they reappear as women: with the reference to 'women wishing to donate to research' (p. 17). Thus, when the preferred subject position of the discourse is inhabited, women are explicitly invoked. However, this specificity is obscured when this subject position might be destabilised. This is exemplified in the reference to 'protecting *people* from the risks of egg donation for research' (p. 1) (our emphasis). In the discussion of the risks or possible problems in egg donation, women become invisible or fragmented.

In relation to therapeutic cloning, women are often represented as agential informed donors of eggs; risk is downplayed or disavowed. When risk is acknowledged, women in their gendered specificity disappear as subjects. However, in relation to reproductive cloning the risks to women are emphasised and women are represented as exploited dupes. Women's imagined desire for cures is rationalised in the case of donating eggs, whilst their imagined desire for children is pathologised in relation to reproductive cloning. A comment from Zavos when we interviewed him about this aspect of the news coverage illustrates the way that women are mobilised as a resource on both sides of this debate:

> Interviewer: one of the specific things that was said in some of the press coverage was that . . . you were putting women's lives at risk because of the kind of – the –
>
> Zavos: They couldn't name one woman. They couldn't name one woman. Okay. Therefore what they're saying is baseless. I've dealt with many women and men, many, many couples. I had a British couple in my office yesterday and I'm expecting another British – it's actually a surgeon that called me yesterday and said 'I want to come and see you, Dr Zavos'. So if they think that they are so stupid to come and see me so I can put their wives at risk, they'd better think again.
>
> (Interview with authors, 9 March 2007)

Women appear as a material resource for science when they are represented as fragmented and objectified bodies. However, women have been invoked to repudiate reproductive cloning, through allegations that they have been passive victims of exploitation (refuted by Zavos above who identifies men as protecting 'their wives'). They appear as agential subjects in dominant cloning discourse only when they are represented as fulfilling the preferred subject position of the willing donor who contributes materially and symbolically to securing the legitimacy of biotechnoscience. At the same time and through the representation of particular gendered bodies, preferred forms of masculinity, femininity and heterosexuality are also secured.

Women as (potential) recipients of cures and reproductive technologies and as carers for recipients of cures

Women, like men, are imagined as recipients for stem-cell cures, as patients or potential patients. This is an imaginary construction articulated for example in Ian Wilmut's speech to the Genomics Forum (Edinburgh, 22 September 2005) when he asked the audience to imagine themselves as future sufferers of motor neurone disease. The representation of individuals and patient groups as in need of stem-cell cures has provided the legitimacy for stem-cell research and been the lynchpin of campaigns for enabling legislation and research funding.[10] In addition, women are also figured as potential beneficiaries of stem-cell research because of their roles as carers, nurturers and custodians of the health of others (as wives and/or mothers). (Clearly not all caregivers are women but the dominant representations are gendered and caregivers are figured as women.) Women rarely appear as spokespersons on these issues, but reference to their needs or desires has become a recurring trope in the discourses of cloning and stem-cell technoscience.

In our examination (in Chapter 3) of the image on the South Korean stamp which celebrated the scientific accomplishments of Hwang and his team, we noted that the narrative is of an apparently disabled male subject who is cured as a result of therapeutic cloning research and who leaps from his wheelchair into the embrace of a apparently female figure. The extensive and widely circulated representations of Christopher Reeve and his 'caring' wife Dana (now also deceased) reiterated a similar iconographic figuring of the male patient and the caring woman seeking a stem-cell cure. When Reeve died, the actor Susannah York, who had been the on-screen 'mother' of his character 'Superman', was interviewed as part of a set of obituary television and internet news items produced by the BBC in early October 2004. In these media productions, Christopher Reeve was positioned as a figure at the centre of a group of women, including his actual wife and his fictional mother, who cared for him. The appearance of caring women around the ailing Reeve became a mainstay of the media representations of him in the latter part of his life and after his death. In relation to his campaigning around stem-cell research, this imagery highlighted women's position as carers who were seeking cures for others and looking to stem-cell research to provide it.

Images of women as caregivers have been more explicitly mobilised by scientists and policy makers in public consultations pertaining to therapeutic cloning in the UK. In the HFEA document 'Donating Eggs for Research: Safeguarding Donors', mentioned previously, the donation of eggs for stem-cell research is presented as an issue that has emerged because of women's demands. This is clear from the title of the document – which presupposes that there will be donation for research. It is also clear from the introduction, which emphasises that 'the purpose of this consultation is to address the question as to whether women should be allowed to donate to research projects' (p. 1). Although there is an explicit acknowledgement later in the document (p. 8) that the issue arose for the HFEA because scientists had been applying for licences, it is *women's* 'right to choose' to donate that is foregrounded in the document, as though this were its driving imperative. It was, in fact, this idea that women had been pushing for therapeutic cloning that provided the legitimisation for Ian Wilmut's suggestion in 2004 that they should be invited to donate eggs for cloning research. This is illustrated by the following comment from Christopher Shaw (Department of Neurology, Institute of Psychiatry) – Ian Wilmut's colleague and the joint holder of the HFEA licence for human cloning with him:

> One reason that we are thinking of using eggs donated specifically for research is that women have approached us to do just that. They felt that they wanted to donate eggs to speed up progress to find new treatments for motor neurone disease. So far the women that have offered their eggs are those whose lives have been touched by the disease affecting a friend or family member.
>
> (Christopher Shaw, Science Media Centre, 26 July 2005)[11]

The mobilisation of women's bodies in cloning discourse can usefully be positioned within a broader reflection about cloning and IVF. The relationship between cloning and IVF involves a complex set of stories about technoscientific developments and practices in both the UK and the USA, explored in greater detail by Karen Throsby and Charis Thompson (Throsby 2004; Thompson 2005). IVF is the 'core technology' (Throsby 2004) for cloning through which women may produce the eggs and embryos required for these experiments. Hence, women who enter the 'ontological choreography' of IVF in order to be made parents (Thompson 2005) may also thereby become experimental research subjects for therapeutic cloning. Because of the high failure rate of IVF, it is likely that, instead of becoming parents through IVF, some of the women undertaking IVF, who give some of their eggs to experimental research, will only become the objects of cloning research.

For much of the twentieth century, prior to the introduction of IVF in the late 1970s, cloning was imagined as the technology which would break the link between women's bodies and reproduction. Of course, both within feminism and outside of it, there has been and there continues to be considerable controversy about whether this is a desirable goal and whether reproductive cloning would be in women's interests. The striking feature of recent cloning discourse is that, in the face of claims that reproductive cloning is possible (and indeed, that it has been undertaken, although these have been discredited), the notion that this is in women's interests has become much more problematic. Hence the representations of women as the recipients of cloning as a method of assisted reproduction have been highly controversial and disruptive to the conventional narratives of paternalistic technological intervention. As IVF has increasingly been constructed as normative and beneficial to women (Throsby 2004; Thompson 2005), reproductive cloning has been disavowed as a mode of new reproductive technology (NRT). However, cloning has been linked to IVF in the visual imagery of recent films. Moreover, this IVF/cloning interface (Franklin 2006) has been intensified and institutionalised by the experimentation at the Newcastle Centre for Life, Ian Wilmut's request for egg 'donation', and the HFEA's licensing and consultation (as outlined above).

For the reasons signalled above, the representation of women as the recipients of cloning as a reproductive technology during the past decade has been complex, often incoherent and always disruptive. In recent media stories about cloning as an NRT in both news and fictional genres, there is always some kind of 'price to pay'. In the UK news stories about Zavos's use of cloning as an NRT in January 2004, his female patients were represented as exploited, pathologised and disempowered through the generic conventions referred to below and through the repeated accusation that he was exploiting vulnerable women. Of course, this is, in some ways, an extension of the familiar trope of the 'desperate infertile' that has circulated in media coverage of other NRTs, particularly IVF (Pfeffer and Woolett 1983; Franklin 1990; Pfeffer 1993).

This trope has been pushed even further in the coverage of Zavos's interventions into the field of reproductive technologies as the women he treated have been represented as not just desperate, but as 'too desperate'. In an earlier BBC documentary about Zavos's work, *Horizon: Cloning the First Human* (2001), the women interviewed were framed as potential perpetrators or victims through the use of voice-overs and disguise or silhouette. These techniques are familiar to audiences as the generic conventions of crime documentaries about unsolved crime, particularly domestic violence, rape and other violent crime.[12] In the documentary about Zavos (2001), news reporting about his claims (2004), and the film *Godsend* (2004), which fictionalises a similar scenario, women appear as much too desperate to find a way to reproduce and the use of cloning as an NRT is shown to have negative outcomes, both caused by and enacted upon women's bodies.

In the *Daily Mail* in January 2004, Dr Zavos's announcement that he had attempted human reproductive cloning was reported in a news article with extensive visual diagrams and models to 'explain' the cloning procedure (see Figure 5.1). Human cloning has also appeared as a reproductive method in some recent film and television drama. However, because of the medicalisation of reproductive technology and the production of IVF as a 'cure' for infertility (see Crowe 1990 for a deconstruction of this normative positioning, as well as Throsby 2004, and Thompson 2005) such representations may be viewed and understood as therapeutic. In this sense, the widespread representation of IVF as a cure for infertility, and the contingent position of babies as a therapy, disturbs the distinction between therapeutic and reproductive cloning that we have highlighted as crucial in the early twenty-first century. It undermines the push to establish that SCNT is about cures *only* and that only *therapeutic* cloning is about cures.

This pattern is evident in the film *The Island* (2005) in which human reproductive clones are grown to provide replacement DNA-matched organs and body parts. This practice is presented as the ultimate in personalised medicine. Thus, within this film narrative, the promise of embryonic stem cells in contemporary cloning discourses – the promise of cures and corporeal regeneration – becomes the rationale for the growing of full-body clones. The cloned bodies facilitate a form of 'regenerative' medicine, through supplying appropriate organs and tissues which can be harvested when required. Moreover, the co-joining of reproduction and therapy is represented graphically when the female clones provide DNA-matched children for childless couples.

This occurs in a visceral scene of life and death as the camera is used to frame the fleshy detail of the childbirth. However, there is no imagery of child/mother bonding after the birth here, since the female clone is killed as soon as she gives birth and the cleaned baby is brought into the waiting room and given to the heterosexual client couple, who bond with the child instead. This scene effects the erasure of the bloody labour of the female body from this sanitised business model of producing body parts and babies to order.

While there has been a proliferation of stories about cloning as an NRT, the negative outcomes attached to reproductive cloning are always related to exceptional contingencies. Thus, the flexibility of these discourses allows for the possibility of some future recuperation of cloning as an acceptable form of NRT. For example, in the open letter to the media of January 2004, written by scientists protesting about the reporting on Zavos (noted above), it was acknowledged that human reproductive cloning would be newsworthy if it proved to be a successful material practice: 'We understand that no journalist can afford to miss the possible story of the first human clone and, if evidence appears to that effect, it will be of huge public interest' (Multiple signatories, 'Open letter to the media', January 21, 2004).[13] The main objections to human reproductive cloning that have appeared recently are that it is currently unsafe and that, as yet, there are insufficiently compelling reasons to risk it. However, 'yet' implies prospects and possibilities, invoking a future when cloning may become a desirable NRT practice. This is also because risk has become a virtually unavoidable consequence of technological developments in late capitalism (Beck 1992). Technoscientific development has generally come to be seen as requiring the weighing up of anticipated risks against imagined benefit. The specific objections to cloning as an ART, which have been articulated by scientists such as Ian Wilmut and Robert Winston, are that it would put the women involved and any cloned children at risk.[14] Hence, at this point, some dominant voices have offered their evaluations of reproductive cloning, but this is by no means a closed book.

Indeed, in recent discourses about human cloning there has been increasing flexibility. Hence, cloning as an NRT already appears as less frightening than it has in the past, less associated with horror, and more closely aligned with IVF. The alignment of reproductive cloning with IVF is particularly significant as, over the course of the last three decades, this process itself has increasingly been normalised as a reproductive process in the Western world (Throsby 2004; Thompson 2005).

Clearly women's bodies appear and disappear in complex ways in discourses about cloning: on the one hand, securing normative meanings, whilst on the other, disrupting them. In independent and niche market films, when cloning is offered as a reproductive technology, there have been more explicitly alternative imaginings than in the more conservative Hollywood texts (such as *The Island* discussed above). *Blueprint* (2003) and *Teknolust* (2002), for example, provide explicitly alternative visions of women-centred cloning stories in which strong women characters actively take up cloning techniques for their own purposes. Although limited in their distribution, the films *Teknolust* and *Blueprint* merit attention because they seem to offer some alternative visions of cloning. In these two films the clones are produced through the 'mother'. Issues of reproduction thus intersect with and structure concerns about identity, corporeality and temporality in these films.

That *Teknolust* produces an alternative set of visual conventions is not surprising. It is a highly self-conscious engagement with technoscience by feminist filmmaker Lynn Hershmann Leeson. The film is both a light and complex take on recent developments in information and communication technologies and biotechnology. This is a film about reproduction and the struggles to control it. It is also an exploration of identity and time. The focus is on the former, with the exploration of the different characters and their desires and needs at the centre of the plot. The female biogeneticist Rosetta Stone[15] creates three self-replicating automatons who are visually figured as clones (all four characters are played by Tilda Swinton). Although Rosetta sees herself as a parent, the clones are her contemporaries and siblings. In the film's conclusion, the clones and their creator effectively live happily ever after in two heterosexual relationships and one lesbian one, all of which are successfully reproductive, albeit, in different ways.

Blueprint is a German/Canadian film in which the female protagonist, Iris, approaches a scientist requesting that he create a clone of her. One of the themes of the film is the question of what constitutes parenthood and family and the narrative includes conflict about the role of the scientist as he attempts to represent himself as a parent. Iris disavows his involvement after the birth and frustrates his attempts to develop a relationship with the cloned child. His efforts to assume a parental role are resisted by both Iris and her cloned daughter Siri later in the film. This, in turn, triggers his anger towards the characters of Iris and Siri, and his eventual announcement to the media about the successful cloning. The mother and daughter are played by the same actor in this film (Franke Potente) and the visual image of the clone as twin is deployed.

The role of the media is emphasised in this film, as much of the early plot revolves around *when* it would be appropriate to disclose the experiment publicly. Unlike *Godsend* in which there has been an agreement amongst the characters that the experiment is secret and never to be disclosed, in *Blueprint* the characters have made a pact that, at some point, the cloning story will be disclosed in the interests of the advancing science. The moment at which the news media are informed about the cloning is the turning point in the film and journalists are portrayed as harassing the cloned child, exposing her to public scrutiny and invasion. The experiment and the early life of the child are dramatically publicised through the intervention of the media. The rest of the film focuses on the cloned child's attempts to take control over her life and establish an independent identity. The metaphors of twin and copy are used to convey the confusion of the character about her own identity. There is a visual reference to the biomedical experimentation by Mengele on the Jews of the Third Reich when she starts wearing a badge, echoing the star of David, with 'clone' on her clothing.

Although this film represents cloning as challenging in terms of the clone's subjectivity, it does not treat cloning as a trope of science fiction or horror. It is a brightly lit film which draws on the conventions of aesthetic realism and naturalistic drama, evoking reproductive cloning as a likely and much more subtle occurrence than any of the action, science-fiction or horror films considered previously in this book. In this film the audience is called upon to believe that cloning is a plausible reproductive choice and to identify with the cloned character as another human. The problems that the clone encounters are familiar to audiences as they concern child/parent conflict and the development of identity and subjectivity.

In *Blueprint* and *Teknolust* these representations of hybrid, postmodern maternity, evoke a complex intersection between bodies and life in which DNA is embodied and cloning appears to be a revised form of maternal reproduction and sibling kinship. Cloning as a reproductive technology involves hybrid kinships through a reworking of sameness and difference. In each case the clones are figured as forms of autonomous life or post-modern humans. In these films parents and scientists are refused creative authority and ownership of reproduction but are materially and symbolically involved in reconfigured reproductive processes and kinship groupings. These different figurations offer alternative ways of seeing cloning and of imagining subjects and social relations around the clone. In these films, the clones are, unusually, not punished through narrative expulsion, death or excess. Overall, these productions provide alternative visions of future/present bioeconomies as they may be re-structured through somatic cell nuclear transfer.

The erasure of women's corporeality

In this section we examine the erasure of the corporeal body. Within recent cloning discourse women's bodies can disrupt the intelligibility of science as rational and beneficial, as well as helping to secure this. However, while the above discussion has highlighted examples of how women appear in cloning discourses, in fact, women's bodies are often simply *absent*. Although we have highlighted examples of women being presented as transparent and naked, in fact they are likely not to be represented at all. In routine discussion of therapeutic cloning (in, for example, early accounts of Hwang announcements of success), women as the source of eggs were simply ignored.

Television news in February 2004 covering Hwang's initial claims featured free-floating images of the blastocyst and enucleation. All the major UK channels – BBC news, Channel 4 news, Channel 5 news – used these images. The Channel 4 evening news covering this opened with a montage of free floating embryos, fetuses and DNA. The voice over announced to viewers that 'an egg is isolated' to explain how the therapeutic cloning story starts. In the same bulletin the only ethical issue considered was the status of the embryo, and issues of consent or exploitation were not considered.[16] This echoes a long history of representation which abstract eggs and embryos from women's bodies, presenting them as free-floating entities (Williams et al. 2003). Science as a global project was emphasised through the appearance of a background graphic of a map of the world whilst the newsreader, Jon Snow, faced the camera. The piece concluded with a quote from the bioethicist Julian Savalescu of Oxford University: 'The lesson from this research is that we need to liberalise our laws across the world on therapeutic cloning.'

Slippage between identifying women as women and simply as 'people' may also obscure gender issues (as illustrated in our earlier discussion of the HFEA consultation document). Women are also written out of the very grammar of the way the process is described. For example readers are told that scientists 'take eggs' or that 'Genetic material is cleared from a donated egg' without reference to women (*The Times*, 12 August 2004). New terminology developed specifically around the *cloned* embryo has developed new forms of corporeal erasure. Terms such as 'embryos not created from fertilisation' and the 'derived embryo' work to dislocate the embryo. Through such designations SCNT embryos appear to have no history: through the negative identification in the first phrase (they are *not* derived from fertilisation) and through the loss or obscuring of context in the use of the term 'derived'. Indeed, the latter begs the questions: derived from where or what? The context of women's bodies and their eggs, from which these new life forms are produced is obscured from the chain of meaning and discursively erased, as SCNT embryos are discursively performed as technoscientific objects. Women are also sidelined by the focus on the ethics around the body/life of the embryo, rather than the body/life of the woman. For example in a story in the *Telegraph* entitled 'This is what the human cloning row is about' (Roger Highfield, *Telegraph*, 16 August 2000), a large image of a SCNT blastocyst took up two-thirds of the story space. The news item was used to explain why 'experts believe the benefits outweigh the ethical considerations' of therapeutic cloning (p. 14).

The erasure of women is also accomplished in, and illustrated by, film. In much Hollywood film, cloning operates as a mechanism for the reconfiguration of reproduction as a masculine scientific process of male parthenogenesis or paternity through which the maternal body disappears altogether (Roof 1996; Battaglia 2001). This figuration operates across a range of representations of different technologies involving information technology and biotechnology (Doane 1990; Springer 1996). In the mass distribution films *The 6th Day, Godsend* and *Blade II*, for example, the processes of reproductive cloning and genetic engineering are displayed as implicated in the displacement of the maternal body. In each of these films, the focus is on the scientist as procreative 'father' in a familiar tale of scientific mastery (Easlea 1987).

In *Godsend,* the mother survives for most of the film but there is no generative force attributed to the maternal body. The scene in which the transfer of the cloned embryo occurs is very indicative in this respect. In this scene the mother is represented as an immobile patient in an operating theatre, who is being scrutinised by imaging technologies, whilst the cloning scientist/doctor transfers the embryo. During the camera shot of the transfer, the woman's body disappears altogether, as the close-up focuses on the scale of the blastocyst. The biological father and the scientist who performs the cloning process are identified visually as the active parents of the child, both at the moment of transfer and later in the film. This is evident when the scientist enrols the father to discuss 'our' child (the scientist's term). The dynamic between the parental scientist and the father is the fulcrum of the knowledge about the child. The maternal body is figured as a passive womb in which the clone gestates and which is subject to the manipulations of the scientist. Moreover, the character of the mother is pathologised as she is identified with excessive grief and emotion. It is the exploitation of this grief by the scientist and her expression of her desire for a cloned child in these terms that is represented as enabling the cloning to occur.

In other mass-market, Hollywood films, such as *The 6th Day*, *Blade II* and *The Island,* the maternal body is absent entirely and laboratory equipment is visually substituted for the womb. These films show the creation of clones through procedures which occur outside of the body and which do not require women. Hence, reproduction becomes a matter of laboratory processes, which are undertaken by male scientists. The cloned offspring are all claimed as the creations of the male scientist in these film narratives, while the female body is either completely absent, invisible, killed-off, or figured as an empty conduit.

The popular film *Mission to Mars* creates an even more striking image of disembodiment through a figuration of DNA as 'life itself'. In this film the focus is on the communication of genetic material between species and it does not represent an embodied clone but figures the genome as a medium. DNA is represented as a universal procreative force sent to 'seed' the blue planet (Earth), invoking analogies of the galaxy as a womb, DNA as spermatozoa and the earth as ovum. Thus, the universe is engaged in a process of sexual reproduction in which sperm meets egg to create life. The focus on sexual reproduction across species through DNA contributes to the development of a narrative in which humanity is figured as always (already) cloned. This is the case because within this plot, viewers are invited to entertain the possibility that humans as a species have been cloned from Martian DNA. This is not cloning as it is usually understood as either individual replication or the mass production of sameness. Rather, cloning occurs here on a much grander scale in which replicating genetic material is portrayed as a fundamental, natural process. This representation plays out the Raelian claim that humans are clones of extra terrestrial beings and that they are thus themselves compelled to pursue the development of cloning technologies.[17] In this trope, cloning is not figured as a discrete set of technoscientific processes, rather, it is portrayed as the fundamental mechanism of species evolution. Cloning as a mode of reproduction becomes crucial and associated with an overarching framework for understanding the development of life. This naturalises cloning – the reproduction of DNA – as the way in which life occurs. This figuring of DNA as the key element in the grand narrative of life resonates with other claims made about the significance of DNA, made in rather different contexts.[18]

The astronauts in the film *Mission to Mars* describe the process of cellular reaction, as the space ship with its DNA hits earth, as 'seeding' Earth. They interpret this to mean that the aliens have thereby become their kin, because of their sharing of DNA. DNA is thus constructed as the core identifying element of both the Martians and the humans and it represents 'life itself'. In this film the 'genetic imaginary' (Franklin 2000) is clearly materialised. The triptych of the blue planet, the foetus and DNA (as seed) that compose the images of 'global nature/culture' as examined by Donna Haraway (Haraway 1997) and Sarah Franklin, Celia Lury and Jackie Stacey (Franklin et al. 2000) are mobilised in this film through the representation of DNA, the blue planet seen from the space ships and the foetal alien 'life' which conforms to the translucent, large-headed, big-eyed figure of alien and foetal imagery.

In this trope DNA is portrayed as a generative universal force, which can be free floating and create life on impact with bodies. As a medium and as information freed from a material base it is consonant with the invocation of male paternity. Both tropes revolve around the dislocation of the female body and the insistence on the power of global masculinity.

Re-stabilising reproduction; re-securing bodies

We opened this chapter by arguing that women's bodies have frequently acted as boundary figures in three different ways which are mobilised at different points in contestations around the meaning of cloning. We have explored how women's bodies may be foregrounded as a resource or as beneficiaries of cures and assisted reproduction. We have also explored how the corporeality of women's bodies may be erased. Through the intersection of these three features of cloning discourse, and through the varied representations of human cloning sketched in this chapter, women's bodies appear and disappear to re-secure cloning technoscience as a legitimate nationally globalised project. They appear as a fragmented and objectified material resource and their erasure reinforces cloning as a story of disembodied rationality. Their reappearance secures cloning as hetero-normative in the face of the danger that cloning could be, or could appear to be, excessively reproductive. Nevertheless, alternative representations do reappear in marginal spaces such as niche market film and art, and emerging, for example, in some of the press releases and media produced by civil society groups such as The Corner House, the Centre for Genetics and Society, Boston Women's Health Collective, Hands Off Our Ovaries and WomenLink.[19] They also emerge against the grain in reconfigurations which explore the potential of human cloning as a form of queer reproduction – through alternative films.

These different figurations can be related to different feminist approaches to biotechnology. One set of feminist critiques of reproductive technologies posits biotechnology as undermining maternal control over reproduction, particularly through the authority of male scientists and doctors. *The 6th Day, Godsend, Mission to Mars* and *Blade II* are films that could be read as providing illustrations, or enactments, of this feminist critical understanding of biotechnology. Another set of feminist arguments contend that biotechnology could help liberate women from corporeal constraints – particularly those associated with pregnancy and childbirth (Firestone 1970). There is some vestige of this argument informing the feminist imaginings of *Teknolust*.

Making direct reference to Ian Wilmut and Robert Winston, one of the stem-cell scientists we interviewed commented that:

> the truth of the matter for me is that reproductive cloning is unsafe at the moment, but eventually, if it's proven to be reasonably safe, I see no reason at all why it can't be used as a form of, you know, as a fertility treatment. It gives a lot more power back to women as well, over their own fertility, and I really have mixed feelings about all these pronouncements from male scientists 'we're never going to have reproductive cloning' and 'it's really bad'.
>
> (anon, interview with Wilmut and Winston)

Her proposal does resonate with Firestone's (1970) argument. However, the idea that cloning could provide a form of reproduction that would liberate women from patriarchal control and sexual reproduction seems to be a marginal one and a rather radical proposition in the context of recent discourses of cloning.[20]

Highlighting these three mobilisations of women's bodies allows us to draw out recent anxieties about maternity, paternity and heterosexuality and to examine how these have played out. Those stories which instantiate paternal reproductive and scientific control offset anxieties about the instability of paternity[21] through the abjection of the maternal body and the reiteration of paternal authority. The stories which reconfigure maternity can be seen as intersecting with fears about the outsourcing of the capacities of maternal bodies through the promise of postmodern maternities. Through the relationality of these terms, these narratives about maternity are also in dialogue with these anxieties about paternity. The story of disembodied DNA rescues a sexually reproductive order from the threat of reconfiguration. In the absence of bodies this narrative speaks to anxieties around reconfigurations of life in a context of databased identities such as digitisation, DNA banks and identity cards (Roof 1996). This disembodied narrative also invests DNA with a spiritual immanence, offsetting the anxiety that identity has become a medium of disembodied information (Kay 2000; Thacker 2004). It re-sacrilises this particular kind of information – DNA – as, in effect, cosmic sperm. This construction closely parallels the metaphor of DNA as operating as the book of life (Kay 2000), which informed Clinton's recuperation of the bioinformatic genome through reference to it as 'God's language' (Nerlich et al. 2002).

These figurations are thus part of a complex pattern of positionings. The installation of conventional gender and sexuality markers may work to offset anxieties, so that new relations can be explored within the safety of familiar reproductive narrative categories. The filmic representations of human reproductive cloning make explicit the subtext of the UK press reporting of therapeutic cloning. This is a subtext of anxiety and promise regarding reproductive cloning. The juxtaposition of representations in these two cultural arenas highlights the work involved in discursively severing therapeutic from reproductive cloning, especially in the UK press but also elsewhere. This is particularly striking since, as outlined in Chapter 2, these two forms of cloning involve the same basic material practice. In our review of these complex discursive patterns and in looking across the various genres we have considered, we have rarely seen women's bodies mobilised in relation to the interests of women. Instead, they are used in mainstream texts to resist epistemic violence in relation to the sanctity of science, to secure its borders. In this discursive (and sometimes material) play and work, women's bodies have generally been mobilised in the interests of public support and funding for genomic scientific research.

However, as has been well argued elsewhere, women's bodies are leaky (Shildrick 1997) and their material-discursive mobilisation produces more than closure. The productive forces at work here do not only operate to validate technoscience as the only grand narrative left, but they have also yielded some unexpected ruptures. Human cloning disrupts 'life' in more ways than one, and these validations provide moments of closure alongside these ruptures, affording opportunities for reformulation. We anticipate that the naturalisations and occlusions in the recent formations of therapeutic cloning will help to secure and extend scientific experimentation in this field. The inequalities around race, gender, sexuality and class that have been part of recent practices of organ and egg traffic, surrogacy, treatment access and vulnerability to experimentation are not addressed in stories about human cloning as securing reproductive futures and/or providing cures.

It is our hope that reframing these stories by locating women's bodies at their centre could help to open the way towards a more explicit address of these issues in debates about developments in cloning and human genomic technoscience more generally. The entanglement in recent discursive manoeuvring, including most obviously the maintenance of the boundaries between therapeutic and reproductive cloning, has obscured some crucial political issues around this field and, in this sense, produced distracting discursive chimeras.

Women's bodies, technologies and feminist frameworks

In examining the relationships between women's bodies and cloning technologies there are tensions between research that highlights the agency of women (Haraway 1997; Thompson 2005) and that which foregrounds structure and constraints on agency (Crowe 1990; Throsby 2004). These tensions have, in many respects, been productive and this may also be the case in regard to feminist engagements with cloning. So, for example, 'ontological choreography' is a concept which has been developed by Charis Thompson (2005) to explain women's positioning in the networks of operations that constitute reproductive technologies. Rayna Rapp's concept of 'moral pioneers' also locates women as agents in technoscientific networks. Structure and agency are also made complex in Donna Haraway and Sarah Franklin's reformulation of nature-cultures (Haraway 1997; Franklin 2000). These reformulations challenge the conventions of nature/culture binaries. Historically, some radical feminisms and ecofeminisms have also emphasised structural oppression, whilst simultaneously reinscribing a technology/nature, male/female epistemological schema (Daly 1978; Showalter 1985).

Recent discourses of human cloning and earlier debates about assisted reproduction both operate in relation to different kinds of 'foetal centredness'. The discourses of new reproductive technologies (NRTs) and assisted reproductive technologies (ARTs) were developed through visual technologies which produced the foetus as a subject (Petchesky 1987; Stabile 1994; Hartouni 1997). The foetus remains central to contemporary cloning discourses, as the visual image of the giant foetus in the *Daily Mail* story examined previously exemplifies (see Figure 5.1). However, the figure of the foetus has generally been supplemented by the embryo which is the key figure at the centre of recent human cloning discourses. Moreover, in such discourses, the embryo appears as a source for stem cells and as such is rendered as regenerative, rather than as a subject in its own right.

Visualisation technologies and techniques similar to those which had been deployed to represent the foetus as a subject are now used to 'show' that the embryo is not an entity but rather, a bundle of cells. The single-cell image of SCNT has taken centre stage in iconic displays which visually deconstruct women's bodies for science. As the story of cloning has been removed from the context of reproduction and become primarily therapeutic, the foetus has generally been cut out of the picture. The foetus has been replaced as the visual centre of reproduction by the image of the cell nuclear transfer, and the nucleus, cell and egg. Where the embryo does appear in factual genres it is framed as the source of hope in relation to the new life of stem-cell cures, not the new life of babies.

One continuity between earlier debates about reproductive technologies (as they were developed in the late 1970s) and more recent cloning discourses (from the late 1990s) is the issue of the gendered power relationships operating around IVF and cloning. Earlier research suggested that the excessive medicalisation of reproduction serves to corral reproductive knowledge as an elite and inaccessible domain (McNeil 1993). These processes render women both literally and figuratively 'out of the picture' (Petchesky 1987; Stabile 1994). This operation of norms through professionalised discourses of reproduction, reiterates patterns of gendered inequality within a heterosexual matrix of power relations (Butler 1990) and gender conservatism (Thompson 2005).

Of practical concern for feminist approaches are ongoing problems in the operationalisation of reproductive technologies. These include inequality of access and cost, and instability of treatment in terms of health hazards, failure and multiple births. Concerns about these issues have accompanied the development of *all* reproductive technologies including contraceptive pills, ultrasound, abortion and IVF. Moreover, these should be seen as contiguous with more recent issues such as embryo selection (for sex or fitness, including ultrasound and preimplantation genetic diagnosis (PGD)) and the uses of so-called 'spare' materials such as eggs and embryos. These issues are crystallised in the current controversies about sourcing eggs for research and in the debates in the UK, the USA and South Korea about nascent commercial egg markets and the allegations of exploitation or unethical egg sourcing.

Finally, the construction of women as the 'desperate infertile' is in operation in human reproductive cloning, as it has been in relation to IVF (Pfeffer and Woolett 1983; Franklin 1990). The images of the infertile or (pre)maternal body have operated to secure IVF as a compassionate and necessary process to alleviate the 'desperation' of the 'infertile'. These constructions currently operate across discursive domains, mobilised by 'maverick' fertility practitioners such as Zavos and others looking to promote cloning as a reproductive technology. This mobilisation of the 'desperate' (pre)maternal subject is also reconfigured in film, television documentary and drama. In the film *Godsend* (2004, USA) for example, it is the loss of the child which is represented as driving the parents, and most specifically the mother, to pursue human reproductive cloning. In the coverage of Zavos, it was his potential clients – infertile women – whom he, and some of the documentary makers who followed his work, represented as driving the development of this technoscience. The figure of the desperately infertile thus leaves open the possibility that human reproductive cloning would be developed to meet this naturalised need. The recent construction that women are pushing the therapeutic cloning agenda, illustrates that the capacity to co-opt the agency and voices of women as the justification for science is highly flexible.

Conclusion

Women's bodies figure in human cloning as the (potential) recipients of cures, carers for recipients of cures, as recipients of assisted reproduction and as a resource for constituting science as a national and global project. Women's bodies figure as normative signs re-stabilising cloning sciences as part of a hetero-patriarchal order and they are made invisible in representations of cloning as a disembodied scientific practice. Through these figurations they are mobilised in the constitution of human cloning as a legitimate project of technoscience, securing the value of cloning technosciences. They are also mobilised both materially and discursively as fragmented and objectified body parts, as wombs, eggs and embryos. Despite women's material centrality to this research, they also disappear entirely in the constitution of cloning as an abstracted, sanitised and disembodied practice.

In the discourses of human cloning which we have traced in this chapter, women operate as boundary figures between acceptable and unacceptable forms of cloning. As boundary figures they help to secure the borders of national identity and respectable science, whilst allowing heterogeneous interpretations and a robust flexibility in discourses of human cloning. In the materials reviewed in this chapter, women's bodies are represented in relation to national difference, to secure the national imaginary, particularly in relation to the promise of biotechnoscience. When women are portrayed as national subjects in the UK they are shown as having agency and choice. They are viewed as driving the agenda of science and as choosing to donate eggs. Hence, when they are asked to donate eggs they can be framed as making informed decisions to support science (or not). However, in the UK context, when women are represented as the subjects of certain other nations, such as South Korea, they generally appear as vulnerable, unable to choose, and exploited – particularly through egg sourcing. When women are represented in the UK press as the research subjects of a UK practitioner such as Alison Murdoch or Ian Wilmut, they appear as able to make choices (such as altruistic egg donation). When they are represented as the research subjects (who are also paying clients) of Panos Zavos, they appear as exploited, vulnerable, gullible and pathologically desperate for babies.

It is evident from these strategic mobilisations that women's bodies operate as a resource for cloning sciences in two ways. First, they constitute a corporeal resource: women's bodies are at the centre of contemporary cloning as their eggs are currently required for cell nuclear transfer. Second, women's bodies are also a crucial resource for securing *meaning* about cloning. At this level, they are used to secure therapeutic cloning as a beneficial scientific practice and simultaneously (but flexibly) to discredit reproductive cloning. They are made invisible, represented as naked and dislocated, and mobilised according to dominant discursive frames for understanding cloning as a rational, disembodied, benign practice which disavows the constitutional centrality of women's bodies.

Whilst women's bodies are mobilised as a resource in making the meaning of science, they may also operate as disruptive forces in the representation of science as a rational and beneficial endeavour and destabilise cloning, nation, sex and sexuality. These disruptions occur in niche market and alternative texts emerging from press releases from civil society groups and in queer and feminist filmmaking (*Teknolust*) and in independent film (*Blueprint*). We have also identified the appearance of women's bodies through some of the sub-textual or inferred codings and readings of more mass-market genres and forms such as the UK press and Hollywood film. We have shown that in the representation of women as the recipients of reproductive technologies and as the carers of the recipients of cures, there have been some disruptive contradictions in the mobilisations of women's bodies. An example of this sub-textual disruption is the simultaneous representation of women as both exploited dupes and as agential subjects in relation to news stories about reproductive cloning and egg donation for therapeutic cloning.

As well as the constitutional contradictions in news media and ambiguity in film, we have also registered explicit attempts to produce alternative imaginings. The films *Teknolust* and *Blueprint* present visions of hybrid and postmodern maternities that decentre hetero-sexualised and patriarchal scripts about nation and science. In these films possibilities for women to be represented as scientists, to celebrate reproductive cloning and to use it to bypass heterosexual reproduction are offered.[22] However, such alternative imaginings are only explicit in fictional texts and in the representation of cloning, as we have noted previously, fiction operates as the figurative 'bad other' in which science is *not* made and in which publics are misled. We shall explore in more detail in Chapter 6 the strategic disavowal of fiction within recent cloning discourses. At present, there is limited economic investment in alternative productions and the opportunities for such alternative imaginings to have an impact on the dominant discourses of cloning are thus limited.

The figuring of women in cloning is embedded in the legacies of previous debates about IVF and embryo research. These debates provide a vital resource for understanding how the relationship between women's bodies and reproductive technologies has developed into the recent conjuncture. Human cloning discourses have been and continue to be formed in relation to these debates. Contradictions between the fear of reproductive cloning and the desire for therapeutic cloning operate through the rhetorical and material mobilisation of women's bodies. These bodies act as discursive pivots for establishing new meanings, and destabilising older meanings about human cloning as it shifts from a fictional (reproductive) trope to a scientific (therapeutic) practice.

6 Truth claims and genres

Introduction

This chapter explores the ways in which scientific claims are *produced* as plausible or implausible, credible or 'incredible'. We considered a range of terms here, including 'crafted', 'manufactured', or simply 'made'. The crux of the matter is that some communicators in some genres assume a transparency of mediation – the scientific facts are real, objective and 'out there'. These facts simply require transmission. However, other communicators are explicitly reflective and self-conscious about the craft that goes into making their claims about science. The terrain becomes even more complicated if we acknowledge that some communicators who carefully craft their claims disavow the work that goes into making them appear credible, whilst what counts as plausible in a Hollywood movie may differ markedly from what counts as plausible in, for example, the speculations of an expert committee convened by a sitting legislature.

We are interested in how some science is (at least temporarily) 'made real', while some is (temporarily) 'made false'. Our particular concern is with recent discussions about human therapeutic and reproductive cloning, but much of our discussion pertains to science more generally. We suggest that claims about therapeutic and reproductive cloning cannot simply and unreflectively be assigned to one side or the other of the divisions between truth and falsehood, reality and imagination, or fact and fiction. We review the flexibility of discourses about therapeutic cloning, an area of scientific research first mooted in 1997, that (as discussed in Chapter 3) initially promised cures just beyond the immediate temporal horizon. We suggest that this is a horizon that has advanced or retreated with reference to key news events. The re-evaluating of the claims of Hwang as fraudulent has been the most obvious and dramatic element in the uncertainty about these horizons. At the same time it is our contention that since 1997, human *reproductive* cloning has come to be presented, by some, as a more imminent technoscientific prospect. This is despite the efforts of other key actors to maintain its positioning as fantastic, indeed, as a science fiction.

Our examination of truth claims about science in this chapter is explored through an explicit focus on questions around diverse media *genres*. In a report about the relationship between scientists and the media, published in 2000, Ian Hargreaves, former Director of the Centre for Journalism Studies at Cardiff University, argues that: 'If we are to understand the way that the media and science interact . . . we need to do so from the premise that crude oversimplification about the media is as damaging as crude oversimplification about science' (Hargreaves and Ferguson 2000: 4). We would agree. We cannot hope here to rehearse the work of an entire and diverse field of research that encompasses dimensions of sociology, politics, cultural and communications studies, to take just a few examples. However, by drawing attention to some of the complex circuitous relationships in play in relation to 'the media' we hope to illuminate some of the nuance that is obscured in statements that characterise a diverse, nationally and regionally specific, multi-genre, multi-platform industry as a single homogeneous object. The media *include,* for example: television, radio, the internet, cinema, regional and national press, specialist trade publications and academic publications. They encompass a wealth of content designed to inform, entertain and persuade. The issues of which channel to select and of which genre or form to choose in order to inform, entertain or persuade – or all of the above – are explored and debated by media professionals as well as many others, including, of course, scientists and policy makers. Some claims made about 'the media' reduce this complexity. Moreover, the lack of attention to detail may obscure crucial features of media operations, including, for example, features of different genres. We aim to avoid some of this confusion, conflation and transposition in our account, whilst demonstrating the rhetorical stakes in moves across channel and genre boundaries.[1]

Alongside challenging generalisations about the media as homogenous, we also think it is important to challenge simplistic dichotomies which involve the assignation of particular media genres or forms to the fixed categories of fact or fiction. We aim to complicate such attributions. In this chapter, we thus develop our arguments about boundary work around human cloning in the media through an exploration of some diverse *genres* involved in producing claims. We reflect on the press and TV news reporting of Hwang's claims (first as true, and then as false) and then we go on to examine some hybrid or liminal media forms. We examine these diverse forms to explore the different generic approaches to stabilising or destabilising the 'epistemic authority' of (the real or fictional) scientist protagonists and the boundary between fact and fiction. We also examine the significance of the *mobilisation* of this boundary in generating and channelling imaginative investment in science (see also Chapter 3 on cloning futures). Furthermore, we explore how fact/fiction distinctions relate to some of the other boundary-making rhetorical practices (especially with regard to therapeutic and reproductive cloning) which we have addressed in this study.

Our three case studies of hybrid or liminal media are as follows:

- The first considers a UK produced television drama-documentary, *If . . . Cloning Could Cure Us* (2004). This drama-documentary involves a fictional scientist in a drama exploring 'real issues' about therapeutic cloning.

- The second focuses on the online marketing activities associated with the film *Godsend*, a Hollywood movie that we have referred to in other chapters of this book (see Chapters 4 and 5). One of the marketing strategies associated with this film involved a website purporting to sell reproductive cloning services.
- The third examines the online promotion of the fertility and reproductive cloning services offered by the real life 'maverick cloner', Panos Zavos (whom we discussed in Chapter 4).

Each of the case studies was initially identified through the monitoring of cloning discourses in the primary mass media. We encountered them first through analysis of our press and television news archive, through the compilation of our filmography of Hollywood cinema and through monitoring television listings related to genomics and cloning. However, as we tracked particular news stories and films through the dominant media forms evoked in discussions about the relationship between 'the media' and science,[2] these primary case studies led us to what, for the purposes of this chapter, we will call secondary media (although such a hierarchy of priorities is clearly arguable in an environment where people take 'multi-media' for granted). In pursuing our analysis beyond the cinema screen or the newspaper page we acknowledge, and indeed insist upon, the permeability of media forms and the diverse sites in which audiences might encounter discussions of human cloning (an issue implicit, but not extensively explored, in our earlier chapters).

Mediating Science facts/fictions

In *Cultural Boundaries of Science* (1999), Thomas Gieryn poses a question which is directly relevant to our research project: '*Where* does the adjudication of the truth or falsity of claims about nature take place?' He proposes extending 'Andrew Abbott's conceptual framework for studying the professions in general to the specific case of science: scientists seek to protect their "jurisdiction" over the "task" of deciding which natural claims are true and which are not' (Gieryn 1999: 186).

Gieryn observes:

> Plainly, most folks on the street will tell you that TV news and congressional meeting rooms are not exactly science. But that commonsense, tacit, and stabilized divorce of science from mass media and from politics is precisely what needs to be explained sociologically.
>
> (ibid.: 187)

Our contention in this book pushes beyond Gieryn's formulation. We propose that, with regard to recent human genomic science – and human cloning, in particular – science *has been made* and *is being made* in the media. In making this claim, we are not denying the importance of laboratory work. However, we contend that the laboratory is not the only site for the making of recent human cloning science and that it is important to widen the focus if we are to understand contemporary technoscience and, more specifically, recent developments in human genomics.

Furthermore, we insist that the adjudication of the truth or falsity of claims about 'nature' is not the *only* issue at stake in relationships between scientists, politicians, the media and the public. The issue of trust is also crucial because of the so-called crisis around public trust in science that has been discussed in various forums, including the UK parliament from the late 1990s. We referred, for example, in Chapter 2, to the establishment of the Science Media Centre in response to the House of Lords Select Committee on Science and Technology Third Report on Science and Society (2000) which suggested that a new culture of open and positive communication with the media was required in order to renew public trust in science. In this chapter, we consider the issue of truth claiming with reference to specific media forms as another, but equally important, context in which these issues have been playing out. Our suggestion here is that the imaginative exploration of technoscience and media portrayals of the way that technoscience will be constituted in the future form a crucial and legitimate part of this landscape. Placed in this context, it may be less important whether or not a Hollywood blockbuster 'gets the science right' in the sense of technical accuracy. The more significant question may be the extent to which a film enables explorations of the social and political dimensions of technoscientific developments.

More fundamentally, this chapter raises questions about the conceptualisation of science itself as a source of unmediated truth. We also question the positioning of the media as a distinct arena which operates quite outside the development of science. Writing about genetics, José van Dijck suggests that:

> Popular representations of science are commonly viewed to be generated by non-scientists – journalists, fiction writers and others. Scientists are generally regarded as 'producers' of scientific knowledge, whereas journalists, novelists or fiction writers are ascribed the function of 'distributing' expert knowledge and creating popular stories. Popular media accounts of science, in line with this model, are viewed as attenuated truths at best or distortions at worst.
>
> (van Dijck 1998: 10)

Van Dijck's analysis suggests that the established view is that knowledge and truth emanate from scientists and that the subsequent process of mediation (through journalism, popular fiction, etc.) generates distortion and/or fiction. Van Dijck's perspective is a critical one and she questions this framing. We regard our research as contributing to these re-appraisals.

One obvious problem with the taken-for-granted model sketched above is that it assumes that truth emerges from the scientist or scientific community in a pure, unmediated form. We have drawn on Shapin and Schaffer's (1985) detailed study of the foundations of modern science to emphasise that the generation of scientific truth is never unmediated and that natural facts themselves are always fabrications of material, social and literary technologies. In this chapter we discuss the ways in which the scientific facts or truths of cloning have been embodied in specific literary technologies, such as the crafting of scientific journals or news accounts, whilst other literary technologies are identified as fictional, and, thus, at best, as technologies for addressing social or emotional truths, and, at worst, culpable of the misrepresentation of science. We also turn our attention to the gold standard in the literary technologies of modern science – the peer review process – to consider the strains which have emerged around it with reference to recent developments in cloning. Our exploration extends to consider specific visualising technologies and the particular form of witnessing that they facilitate. With regard to new information technologies, in our discussions of key websites, we will touch on issues of production values and editorial control in the recent communications environment in which producers of facts/fictions have had a proliferating number of platforms on which to mount their messages.

Simon has shown that, in addition to its taken-for-granted role in mediating science to the lay public, in periods of controversy the mass media can also act as a channel for inter-specialist communication (Simon 2001). His account of the relationship between the scientific community and the media, which draws on earlier work by Bruce Lewenstein and Massimiano Bucchi, undercuts the vision of science being continually embattled by the media in ways that undermine science or lead to public misunderstandings of it. Simon insists upon 'the importance of science in public as a kind of *public science*' (p. 387). He provides an outline of how he sees this operating:

> In public science, scientists work to construct or deconstruct knowledge in the public domain rather than just within specialist networks, and their work is mediated through mass media forums like newspapers, magazines and television rather than just through the apparatus of scholarly publication and presentation . . . public representations of science can produce knowledge as well as culture.
>
> (p. 387)

In addition to highlighting the public making of genomic science, throughout our analysis we have pointed out the transpositions, borrowings, exchanges and flows between factual and fictional sites which have contributed to the cultural imaginings of cloning. Our contention is that these have been crucial in the making of recent cloning technoscience. As this suggests, fact and fiction are much more interrelated than the model van Dijck scrutinises would suggest. For example, in *Primate Visions* Donna Haraway points out that: 'Facts ought to be discovered, not made or constructed' (Haraway 1989: 3). Yet, she observes that the etymology of the term 'fact' 'refers us to human action, performance, indeed to human feats', facts are that which is 'known by direct experience, by testimony, and by interrogation' (ibid.). Haraway's probing thereby demonstrates the constructed nature of both 'fact' and 'fiction'.

Christopher Frayling's *Mad, Bad and Dangerous? The Scientist and the Cinema* is a rather different project from Haraway's *Primate Visions*. Nevertheless, there are some affinities between these two texts. In his tracing of various genealogies of the cinematic figure of the scientist, Frayling unearths a complex pattern of interactions and exchanges between fictional films and 'factual' science and technology.[3] Frayling does not himself offer a full-blown analysis of this pattern (and his sketch is somewhat unevenly filled out). Nevertheless, his study does gesture towards a much more complicated assemblage of processes than the somewhat reductive model outlined above or that offered in the crude sender – (message) – receiver model of the communication of scientific truth familiar from some Public Understanding of Science discourses (Frayling 2005).

Jon Turney is another researcher who has investigated the terrain of science and fiction. Turney offers the valuable observation that: 'fictional representations matter, that the science and technology we ultimately see are partly shaped by the images of the work which exist outside the confines of the laboratory report or the scientific paper' (Turney 1998). This is a crucial insight, but we believe that the disaggregating of fact and fiction may not be as easy as Turney's framing of this relationship seems to indicate; nor is it necessarily desirable. The achievement of Turney's work lies in his detailed analysis of the significance of *Frankenstein* as the 'governing myth of modern biology' (ibid.: 3). In this sense, we have taken inspiration from his determination to study the making of the biosciences in recent Western popular culture.

Cloning claims in the news

News coverage is generally considered to be the ultimate vehicle for delivering the truth in the media. There is an expectation that news coverage will tell us what has happened and that it operates in the interests of providing truthful information. For this reason, journalists often draw on official or accredited sources. Where their sources are not accredited or identified with particular positions or views, it is conventional to expect that journalists should and would alert their audiences. The news reporter is supposed to be more 'watchdog' than 'lapdog' and, therefore, wary of manipulation and spin. Particular established media institutions have built their reputations on the accuracy of their news coverage. So, for example, BBC TV and *The New York Times* have gained or staked their reputations on their ability to provide an accurate 'record' of events and facts, as well as presenting interesting commentary and opinions which are clearly designated as distinct from the record of events.

Thus, news reporting is often presented as a genre with a privileged relation to truth. For this reason, it is generally distinguished from other media forms that are regarded as primarily fictional or unreliable. Hence, news coverage is often contrasted with Hollywood films because it is concerned with fact not fiction, with truth rather than imagination. Likewise, in the provision of information, quality news reporting has come to be distinguished from unregulated websites produced by groups and individuals in that such reporting is associated with validity and accountability. The news media's commitment to accuracy,[4] combined with the professional skills and expertise of media personnel, are seen as delivering products that are in marked contrast to those generated in other media contexts.

The high expectations about journalism aligning with truth undoubtedly contributed to the expectation articulated in the 'Open Letter to News Editors', issued under the aegis of the UK Science Media Centre (previously discussed in Chapter 4), that news journalists should filter out or contextualise mavericks such as Zavos (Science Media Centre 2004). Indeed, more generally, Zavos's dealings with the news media have been characterised as self-publicising, with this form of public communication being likened to marketing or advertising. For example, *The Sunday Times* referred to him as a 'modern-day scientific snake-oil salesmen' (Editorial 2004a) while the *Independent* suggested that: 'He is thought to be coming to Britain to generate further publicity for reproductive cloning' (Laurance 2004). Such characterisation either explicitly or implicitly invokes a model of reputable scientists' public communications as selfless and as serving an educational purpose in a democratising mode. The critical perceptions of Zavos indicate the circulation of expectations about how scientists can and should interact with the news media (see Haran 2007 for further discussion).

However, while news values are supposed to ensure objectivity in the reporting of facts, recent coverage of technoscience in the UK has been rather different. Recent technoscientific stories linked to human health applications have frequently been framed through emotive headlines that revolve around putative threats to normative assumptions about social life. For example, in 2004, researchers at the University of Newcastle applied for an HFEA licence to undertake research with embryos on mitochondrial DNA that would involve removing the nucleus from an embryo created through IVF and replacing it in a donor egg. Responses to this licence application included the *Daily Mail* headline: 'Science Seeks to Deliver a Baby with Three Parents' (Editorial 2004b). When the licence was eventually awarded many months later, even the so-called 'quality press' rhetorically conflated embryos with babies. In the *Independent*, the headline read: 'Test-Tube Baby Will Have Father And Two Mothers' in the first edition (Connor 2005a), and 'Scientists Given Right To Create Baby With Two Genetic Mothers' in the final edition (Connor 2005b). There was even more hyperbole in the front-page headline of the *Telegraph*, but it reworked a somewhat tired trope: 'Designer Babies To Wipe Out Diseases Approved' (Highfield 2005b).

Returning to the reporting on cloning in particular, it seems obvious that 'the news' is about much more than providing 'the truth'. The attention given to Zavos's announcement of his cloning achievements in January 2004 is an obvious case in point. Why did UK television news bulletins on 17 January 2004 and UK newspapers on 18 and 19 January 2004 feature a story about human reproductive cloning announced by a man described in such uniformly pejorative terms – for example, 'extraordinarily irresponsible', 'a maverick and egotistical doctor' – as detailed in Chapter 4 (Gledhill 2004; Jones 2004). The scientists who were outraged by both Zavos's claims and the coverage they were given in the UK media offered one rationale: 'We understand that no journalist can afford to miss the possible story of the first human clone' but they immediately followed this with the caveat: 'if evidence appears to that effect, it will be of huge public interest' (Science Media Centre 2004). Hence, both the reporting of Zavos's claims and the letter of protest about UK media coverage of his announcement issued by some of his scientific opponents employed a similar rhetoric. Attention is drawn to the values that scientists and journalists are purported to share: organised scepticism and rigor in providing evidence to substantiate claims. Zavos's claims were consistently represented as suspect, although one scientist who was quoted did assess that Zavos had the expertise necessary to perform reproductive cloning (Robin Lovell-Badge on Channel 4 evening news, 17 January 2004). In fact, the truth status of claims about cloning was specifically linked, by both journalists and scientist sources, to publication in peer-reviewed journals, in advance of press conferences. The use of high-status, peer-reviewed journals to provide the source material for science stories in the news is normal practice, documented extensively in academic articles and books on the public communication of science. This conventional practice was also further publicised in the popular accounts of the press relations surrounding the announcement of the birth of Dolly the sheep (see Kolata 1998; Wilmut and Highfield 2006; Chapter 2 this book).

On Channel 4 News in the UK (evening bulletin, Saturday 17 January 2004), Alex Thomson took up the cudgels on behalf of the truth. After an introductory piece setting out Zavos's claims, Thomson conducted an extremely combative interview which left the viewers in no doubt about the scorn with which the journalist viewed his interviewee. By repeatedly suggesting that Zavos's behaviour was not that of a typical scientist, since 'scientists simply don't behave like that', he marked Zavos as an outsider to the legitimate scientific community. Thomson asserted that there was not a shred of evidence for Zavos's claims because the scientist had not published in peer-reviewed journals. Zavos countered that he has published over 800 articles, but Thomson immediately returned with the query: 'Peer-reviewed?' Interestingly, Zavos pointed to the publication cycle of peer-reviewed journals as an obstacle in his claims-making. He maintained that, if Thomson knew anything about the process of scientific publishing, he would appreciate that any submitted article would not be published swiftly enough to establish his claims to priority if a successful pregnancy were to be achieved from the embryo implantation. That is, the peer-review process is so lengthy that a scientific article might not achieve publication in advance of the successful outcome of such a pregnancy, and that might pose difficulties with regard to the balance between making 'the news' and establishing his scientific credentials. On this television news bulletin and in many of the UK press news stories about Zavos's claims, it was repeatedly suggested that his failure to publish his work in peer-reviewed journals was sufficient grounds to undermine his status as a credible witness. This was despite the fact that, as Zavos himself indicated, there had been insufficient time for a paper on this research to have been submitted, peer-reviewed and published. Indeed, many months later (July 2006), Zavos did publish an article based on this work in *Archives of Andrology* – a journal which David Adam of the *Guardian* described as 'a little-known specialist journal' (Adam 2006). We discuss Zavos's publication strategy further as part of our analysis of his website later in this chapter.

Peer-reviewed scientific journals are a crucial part of the literary technology of contemporary science which derives from Boyle's original model of experimental natural philosophy. The disruption that is occasioned when scientists circumvent the established route of publication in peer-reviewed journals, followed by press releases which result in popular accounts of their work in press or television news has been analysed in depth (Lewenstein 1995; Gieryn 1999; Simon 2002). However, as was the case with Boyle's original literary technology, peer-reviewed journals gain their credibility from the social technology in which they are embedded. The networks within which draft articles circulate and the status hierarchy of journals form part of this social technology. These extra-textual factors are also crucial for readers: they inform their expectations regarding such publications and their reading of them. So, for example, publication in *Nature* or *Science* signifies that an author is considered to be a highly credible scientific witness by their peers. All genres operate through the mobilisation of expectations regarding the form of texts and through such extra-textual factors, but with literature that carries the imprimatur of 'Science' this is even more pronounced (Stern 2004).

However, despite the enormous investment in it, the peer-review process does not necessarily guarantee truth. The document produced by the London Science Media Centre, 'Peer Review in a Nutshell' notes, 'Peer review cannot pick up certain types of misconduct. If someone is deliberately cheating then they can get through the peer review process' (Science Media Centre 2003). Unfortunately, this caveat was not one that was widely acknowledged in recent cloning news coverage, either of the work of designated mavericks such as Zavos or of globally renowned therapeutic cloners like Hwang. This was the case until the Hwang scandal provided – literally – graphic proof of the limitations of the peer-review process.

We have sketched the key moments of Hwang's fall from grace in Chapter 4. In this chapter, we consider the strain this case has put on the reputation of the process of peer review and of peer-reviewed journals. We also show how the processes of mediation which had brought Hwang's team's work to international attention were rendered highly visible once the scandal broke. This self-consciousness about mediation within the media was in marked contrast to their earlier reporting on the papers published in *Science*. In summary, in the Hwang case, the legitimacy of the mass media acting as a conduit for 'second-order virtuality' (Gieryn 1999), as outlined in Chapter 2, was initially embraced and subsequently invalidated in the wake of the discrediting of Hwang's work. Indeed, through some explicit interventions into the media, the scientific community – defined broadly to include enabling regulatory bodies and legislators – took credit for expelling this wrong-doer and apportioned additional blame to the mass media for contributing to the scandal. However, there was never any acknowledgement that this additional layer of witnessing – provided by the media attention originally given to Hwang – itself played a significant role in constructing an international community of witnesses who could scrutinise Hwang's claims and ethical practices. This blame apportionment did not go uncontested. In fact, some media journalists have criticised the international stem-cell research community for being so slow in investigating the concerns that had emerged about Hwang and his laboratory, prior to the full-blown global eruption of the scandal. Bart Simon surmises, with reference to a rather different techno-science controversy, that 'without the early involvement of the media cold fusion may have remained a legitimate scientific problem for much longer than it did' (Simon 2001: 388). We could surmise similarly that, without the global media attention initially given to Hwang's research, it might have taken a great deal longer to debunk it.

Initial validation by peer review

During the reporting of the 2004 and 2005 breakthroughs, the factual validity of the Korean work was repeatedly underlined. UK scientists were quoted as explicitly confirming that, unlike previous claims made by 'bogus' or 'maverick' scientists such as Zavos, this work had been conducted by a reputable team and had been subject to peer review (Connor and Arthur 2004; Utton 2004; Wheldon 2004). The tenor of the reporting was clearly affirming. The UK media assessed that Hwang and his team had complied with established scientific procedures and explicit references were made to the peer-review system. So, for example, one journalist noted that: 'Independent referees have confirmed [the work]' (Henderson and Hawkes 2004). Moreover, there was a consistent and recurring playing-off of Hwang and his team against 'rogues' or mavericks such as Zavos. Hence, the BBC Radio 4 listeners were assured that: 'This was quite a different case from Zavos' (*Today*, BBC Radio 4, 12 February 2004). Readers of *The Times* were told that Hwang and his team were not 'rogue doctors, 'cowboy cloners' or 'Karate Kids' (Henderson and Hawkes 2004).

Thomas Gieryn's account of the management of the press conference for the first announcement of the alleged discovery of cold fusion at the University of Utah is suggestive for considering the importance given to press interest in

Hwang's work (Gieryn 1999). Gieryn comments that because the research that was reported was not recognised as being normal or established science, the press conference was:

> in effect a invitation to the media to join Pons, Fleischmann, Brophy, and Peterson in *making* the scientific breakthrough – to become an instrumental and necessary part . . . The media were asked to add their own instruments (videocameras, tape recorders, images, the printed word) to the technical equipment (calorimeters, scintillation counters, palladium rods) that had carried Pons and Fleischmann up to this exciting moment but could carry them no further. The press conference became part of science, as journalists were enlisted as vital allies in making a scientific discovery . . . The media here became the late-twentieth-century equivalent of what Steven Shapin has described as the 'gentlemen witnesses' so vital for attesting knowledge claims three centuries earlier.
>
> (Gieryn 1999: 200)

There are some notable distinctions between the emergence of cold fusion and recent developments around therapeutic cloning and stem-cell research. Perhaps the most significant of these is the involvement of a larger community of scientists and, indeed, of legislators and potential end-users, than was the case with cold fusion in the late 1980s. Nevertheless, there are striking similarities between these two controversial fields of technoscience in terms of the role of the media in their making. The news events of February 2004 and May 2005 'became part of science' in that, in somewhat different ways, they were opportunities for international scientists to enrol publics in support for their research. The publication of Hwang's papers in *Science* – a peer-reviewed journal with a global reputation – was linked to South Korea's desire to be taken seriously as a centre for world-leading bioscientific research. In addition, the availability of photographic evidence that could be used both to substantiate the claims and to illustrate the technoscientific achievements to a wider audience seems to have over-determined the newsworthiness of the events of both February 2004 and May 2005.

Hierarchy of genres

One of the key ways in which Hwang's loss of credibility was signalled was through reference to a range of fictional genres. *Nature* referred to: 'The sheer Shakespearian drama of the Korean cell biologist's eclipse, surrounded by fawning courtiers and plotting groups of acolytes and enemies and in full view of the television cameras' (*Nature* 2006). This framing of the case connoted high drama. In the *Telegraph* Roger Highfield knocked Hwang's story down the generic pecking order when he characterised the events as 'a scientific soap opera', casting the would-be gentleman witness as a character in a feminised popular media form. Even when Highfield employs the term 'saga' with reference to the Hwang case, this seems to designate only the long-running character of the narrative, not a heroic epic (Highfield 2005a). The *Korea Times* drew a graphic and extensively referenced picture of how Hwang's media relations were implicated in his downfall:

> Hwang's menagerie of cloned creations . . . were presented to the goggling press and general public with all the media wizardry and showmanship of filmmaker Carl Denham (Jack Black) in 'King Kong' (2005) or theme park entrepreneur John Hammond (Richard Attenborough) in 'Jurassic Park' (1993) . . . As Hwang did with his biotech achievement, both film characters capture our collective imagination and dazzle us with jaw dropping, eye-popping scientific stunts worthy of the silver screen . . . When a great accomplishment is twisted into spectacle, it invariably turns against the obsessed visionaries that conjured them.
>
> (Iglauer 2005)

This vivid conjuring of Hwang as media showman and the reference to capturing 'our collective imagination' implies deception and fiction. Hwang is associated with popular film characters who entertain and dupe the public through their productions of fictions (film or theme park). The double layers of referencing of media (film scenes which portray film-makers or theme-park designers) deepens the charge against the discredited scientist. Moreover, it sets the play of the imagination in sharp relief with the production of truth. News reports are about the production of truth, while film texts are about the production of fiction. Furthermore, this obscures the ways in which news stories about therapeutic cloning also call upon our imaginations through the promise of future cures and the relief of suffering, as discussed in Chapter 3. Indeed, the whole technology of 'virtual witnessing' instantiated in peer review itself depends on an imaginative reconstruction of reported findings.

In the wake of the Hwang scandal, the peer-review system of scientific journals has been recognised as no guarantee of authenticity. However, another intervention in the UK press attempted to rescue this established literary technology of science and to re-establish the credibility of the scientific community and the mass media's ability to generate truth. The editor of the prestigious medical journal the *Lancet*, Richard Horton, responded to the media coverage of the Hwang scandal with an article that appeared on the leader page of the *Guardian,* suggesting that re-establishing the credibility of the peer-review process should be the key priority for the scientific community. Horton optimistically observed that: 'The lesson nobody has drawn from the cloning fraud is that science has succeeded not failed. Scientists have quickly rooted out a fabrication of staggering proportions, a self-correction which is to science's credit, not shame' (Horton 2006: 39). In this rhetorical move, science was idealised as the 'area of society [that] promotes such persistent self-criticism, acknowledges its errors so transparently, and rewrites its record'. Horton contended that: 'The public should feel confident that science is able to admit its mistakes and clean up its act.' Horton also staunchly defended the autonomy of scientists and science, arguing vehemently against more regulation, whilst acknowledging the need to ensure that 'the way science is done' is 'more widely appreciated'. There was an intriguing tension at the heart of Horton's article in that he advocated that science is best done 'slowly, quietly and progressively' and yet he also expressed his wish that scientists and science journalists would make greater efforts to ensure that the public could understand how science really works. Horton's conclusion that 'the success of science in our culture depends on a bond of trust between scientists and the public' serves to marginalise the public (more on this in the following chapter) and to restrict scientific witnessing to the confines of the scientific community. It is then rather unclear what would be left for science journalists in the mass media to report on, despite the paradoxical importance of second-order virtuality evidenced by this article. Horton's emphasis on provisionality, corroboration and replication seems at odds with current versions of 'science news'. It would also seem to require journals like *Nature* and *Science* to cease their practices of issuing press releases about scientific papers that they publish, or else to frame these far more cautiously than is the current practice. As the issuing of such press releases has become an integral part of global scientific practice and academic publishing, it seems unlikely that this practice will be altered. In linking another fraud story to the Hwang case, Mark Henderson, science editor of *The Times* took a similar angle to Horton. Commenting on peer review he declared:

> Referees can and do challenge methods and interpretations, but it is hard for them to detect a meticulous fraud. They must, to an extent, take on trust the raw data on which a paper's conclusions are based. Innocent mistakes sometimes slip through, and clever manipulation is fiendishly difficult to spot. What happens next, however, raises an almost insurmountable bulwark against fraud and demonstrates the peculiar rigour of science. A published paper must contain all the data on which its conclusions rest, and the protocols needed to repeat the experiments. It will be pored over by experts across the world, many of whom will try to replicate it. Any errors, whether accidental or fraudulent, will come to light under such intensive scrutiny. When the results of an experiment prove impossible to repeat, alarm bells ring.
>
> (Henderson 2006b)

Henderson's description of 'what happens next' suggests that scrutiny of and attempted replication of the original peer-reviewed paper is what will uncover accidental or fraudulent errors. This obscures the rest of the web of science communication, including crucially science news in the mass media, that might draw the attention of diverse experts and of concerned members of the public (Lewenstein 1995). It also assumes that the papers in question in their own right will necessarily elicit sufficient interest to open them up to scrutiny. In the Hwang case, the international dissemination of the key findings of the *Science* papers through mass media coverage afforded this possibility. A less widely publicised fraudulent paper might well have remained unscrutinised for much longer.

Hype and the management of expectations

The narrative emerging is of a paradox at the heart of the media production of science whereby, in order to obtain the funding necessary to conduct research, scientists and scientific institutions are constrained to make performative claims about the likely applications of their hoped-for results. They are then in the business of managing expectations. Indeed, some versions of the Hwang story suggest that the fraudulent data that he published documented results he fully expected to achieve in the imminent future, but that the superhero of science status that had been bestowed on him in his home country meant that he felt compelled to deliver prematurely. Hwang seems to have been hoisted on his own petard of 'discursive overbidding'.

Discursive overbidding is a rhetorical practice identified by Nerlich and Clarke as operating in media debates about cloning in 2001, occasioned by repeated claims by Panos Zavos and Severino Antinori that they were about to clone human babies (Nerlich and Clarke 2003). It describes the process whereby: 'scientists talk up the potential value of work for which they are seeking the support of over-subscribed funding agencies', or indeed, we would add, for which they seek public approval or a permissive regulatory apparatus (Hargreaves and Ferguson 2000: 16). Therapeutic cloning certainly seems to have been subject to discursive overbidding, but the close of 2006 saw some interventions from key UK scientists in the field of stem-cell research that aimed to counter this.

According to Mark Henderson of *The Times*: 'The medical promise of therapeutic cloning has been oversold and its unreasonably high profile risks turning the public against more promising aspects of stem-cell research, according to one of Britain's most respected experts in the field.' Henderson went on to note that Professor Austin Smith of the University of Cambridge had also remarked that 'cloning research "clearly upsets the general public" yet it has limited potential for treating disease and adds little to scientific understanding of human biology' (Henderson 2006a). Austin Smith moved from the University of Edinburgh to Cambridge to be the Director of the new Wellcome Trust Centre for Stem Cell Research in Cambridge which officially opened on 18 December 2006. His successor at Edinburgh is Ian Wilmut, who as we have noted previously, has been one of the key proponents of therapeutic cloning.

Stephen Minger also spoke out about therapeutic cloning in the context of the news that the HFEA had granted a licence for altruistic egg donation for therapeutic cloning to the Newcastle Centre for Life prior to the completion of its public consultation on this practice. Apart from remarking that the timing of the HFEA decision seemed improper, he stressed that there was not yet sufficient evidence of the benefits of therapeutic cloning to offset the risks it posed to would-be donors of eggs (*Today*, BBC Radio 4, 21 December 2006). He framed the potential of this area of research rather more positively, however, in early 2007 when he was one of a number of UK scientists urging the HFEA to resist government pressure to deny licences to perform somatic cell nuclear transfer by inserting human donor nuclei into the eggs of non-human animals.

While there has been, to use Nerlich and Clarke's (2003) terms, notable 'discursive overbidding' around recent genomic and cloning science, we would emphasise that there are further dimensions to scientists' claims and counter-claims in the media. As Nerlich and Clarke demonstrate, funding has been a factor here. However, garnering support for facilitating regulatory structures has also influenced this pattern. Reputations may be and have been made and destroyed through such claims-making. More generally, it is also important to register the rhetorical dimensions of such interventions.

In this book, so far we have shown how stem-cell research using cloned human embryos is surrounded by the language of promise. Hope is built on claims both about what has been delivered so far and what the future will bring. Such promotion attracts investment – both financial and legislative. The Hwang 'breakthoughs' served initially to justify optimism, verified by the practice of peer review. The subsequent scandals temporarily destabilised claims made for the cures that might be delivered by human cloning. However, some scientists, and some science journalists, rapidly mobilised to re-establish Science's claims on establishing 'truth', and the media became the scapegoat for at least part of the problem. Hwang's playing to the public gallery and superstar status were emphasised in the context of discrediting him.

Having reviewed how claims may be established as plausible, or implausible in the mainstream science and the mainstream news media, we now turn to our three case studies of liminal media.

The drama documentary *If* . . .

Our first case study concerns a drama-documentary or 'dramadoc' aired on the BBC as part of a series called *If*. . . . The centrality of therapeutic cloning to imagined futures in the UK context can be inferred from its choice as one of the topics addressed in this series. This series was produced by BBC television and broadcast on the BBC2 channel. On the BBC's website, the series is described as follows:

> *If* explores and analyses the big issues facing us in the years ahead. The programmes are drama documentaries which create future scenarios based on existing trends and technological advances. The scenarios are fictional, the interviews and the issues raised are real and matter now.

Other programmes that appeared in the innovative series include: *If . . . The Oil Runs Out* (30 May 2006) and *If . . . We Stop Giving Aid to Africa* (26 June 2006). These dramadocs were produced within the Current Affairs division of the BBC, which deals with news-related programming. However, these programmes also employed dramatic conventions and they thus involved different production and screening conventions from news broadcasts. In *No Other Way to Tell It*, Derek Paget traces the converging histories of the emergence of the drama-documentary in the UK and the docudrama in the USA, characterising the former as 'more solemn' and the latter as 'more entertainment-led' (Paget 1998: 6). Having suggested that there are two traditions, he frequently fails to distinguish between them, but we would draw out two key points that he makes about their combination of 'both a documentary and a dramatic provenance' (p. 1). First that this produces 'an essentially transgressive form' because of its resistance to either/or assignations of truth or fiction (dramatisation); and second, that 'the ultimate aspiration' of the dramadoc 'is to make a difference in the historical and political world beyond the television screen by going to places that are originally denied to the camera' (p. 10). Both these points apply to *If . . .*, but in rather different ways to the dramatic reconstructions of historical or contemporary events that Paget discusses. *If . . .* is a speculative dramadoc, and as such it evokes the extrapolative tendencies of science fiction whilst striving to avoid any of the visual conventions associated with contemporary science fiction film.

If . . . Cloning Could Cure Us featured a fictional near-future scenario in which a young woman scientist, 'Dr Alex Douglas', is on trial for breaching the terms of the Human Fertilisation and Embryology Act. The specific breach with which she was charged was allowing embryos created through somatic cell nuclear replacement to develop in the laboratory beyond the fourteen-day cut-off point mandated by this Act. The drama centres mainly on the courtroom arguments, with a few scenes set outside the courtroom. The other scenes take place at the home of a young man with spinal injury, 'Andrew Holland', who the scientist – who was also a medical doctor – hoped might be cured through the provision of patient-specific stem cells that would repair his spinal cord damage.

In addition to the dramatic scenes, the programme also included testimony from a range of interviewees who explained or argued about the scientific, ethical and legal implications of this case. The interviewer remained off-camera and did not speak. The interviewees included some of the key figures in recent debates about cloning and stem-cell research in the UK who were mentioned in Chapter 2, including Suzi Leather, the then Chair of the HFEA, and Josephine Quintavalle of CORE. The programme concluded with the on-screen jury retiring to consider their verdict. Viewers of the programme were invited to cast votes indicating whether they thought the scientist on trial was guilty or innocent. On the basis of these votes, one of two alternative endings would be screened.

The premise on which viewers were voting was one of the least effective parts of this drama-documentary. The scientist did not deny that she had breached the Act. The only defence offered was a 'defence of necessity', which involves a fine legal point that was not fully explored in the programme. The fictional barrister drew an analogy with a case of abortion performed by a doctor prior to the decriminalisation of such a procedure by the UK Abortion Act of 1967. She explained that the abortion had been performed to save the life of the pregnant woman. However, Dr Thomas's breach of the law could only be construed as saving Andrew's life, if the point was stretched to encompass the probability that his life would be shorter with spinal damage than without it. (This limitation to his lifespan was mentioned, but again not fully explored in the programme.) Hence, given this particular legal framing, it would seem as if the only verdict that the jury could return was a guilty one. Of course, construing Dr Thomas's intervention as a matter of life or death identifies therapeutic cloning and stem-cell research with both cure delivery and the promise of universal able-bodiedness. This construction, which we would argue would be familiar to many viewers from other sources than the immediate text, may account in part for the fact that when put to a public vote, the viewing audience voted overwhelmingly (81 per cent) that the defendant was 'not guilty' and the 'not guilty' ending was screened. A final twist was that in this screening, Dr Thomas (a US citizen) is shown being arrested immediately upon her acquittal. Viewers are told that she faces extradition to the USA for breaching that country's cloning laws.

If . . . Cloning Could Cure Us is intriguing because it brings together 'real issues' and 'fictional scenarios' in a manner that provides its viewers with factual information to encourage them to make moral or ethical judgements. The programme scheduling process was also complex since the screening of the drama-documentary had to be split into two segments to enable the audience to select their preferred ending. In addition, a special *Newsnight* studio debate around the issues raised in the programme offered 'more context and analysis' (Downes 2004). This prestigious current affairs programme gave further legitimacy to the dramatic treatment of therapeutic cloning. As with the social technology that underwrites the reliability of peer-reviewed publication, the reputation of *Newsnight* authorises this innovative genre. According to the editor of *Newsnight,* the mission of BBC2's flagship current affairs programme is: 'to make sense of the day's news, to try to explain the detail of current events and hold to account those responsible for them. To make you think again.' He also notes the flexibility of its format, which he argues captures the nuance of current affairs. This is in contrast with conventional news broadcasts which package news items in segments generally between one and four minutes in length; on *Newsnight* a single item can last anywhere from 30 seconds to 49 minutes giving its producers far more freedom to pursue an item in depth if so desired (Barron 2005).

The *If . . .* series, like *Newsnight,* was a television programme with a mission. With production costs underwritten by public money and no obligation to sell advertising contiguous to its screening, its producers thus had the liberty to be more experimental with form, more wide-ranging in content and to deliver a smaller audience than would be possible in a commercial channel. So, for example, this *If. . .* juxtaposed the promises of therapeutic cloning with the risks posed to women egg-donors who might suffer to provide the material such experimentation requires. It also successfully dramatised some of the motivations for pushing scientific boundaries in this field, including the desire to deliver cures and ambitions to undertake cutting-edge research. These and other features of this programme meant that it provided a fuller picture of the social and political issues around human genomic technoscience in the early twenty-first century than is usually provided in a news story of thirty-seconds on television or in a few hundred words in a newspaper. Despite these achievements, as suggested above, the binary voting schema rather closed down the exploration, although the surprise element of the extradition storyline did open up the issues again. This fictional future scenario, true to its mission, remained strongly grounded in contemporary discourses about cloning, extrapolating only very slightly into the imagined future and resolutely steering clear of fantasy or fantastic tropes or plot elements. The cloning story in our next case study, however, seems to be entirely unclear about its narrative trajectory or its genre status.

The *Godsend* film and the viral marketing website

Godsend, already discussed in previous chapters, is what some commentators would regard as the stereotypically 'bad' film which is likely to make the public inappropriately fearful and distrustful of science and scientists. For example, the *San Francisco Chronicle* reported about the film under the headline: 'Hollywood takes a look at cloning – and opens up a can of worms'. In this article, David Ewing Duncan offered the following striking appraisal of *Godsend* and its website (which we discuss below):

> Unfortunately, both the website and the film play fast and loose with the facts – when they don't have to. And both fail to challenge us to think clearly about the subtleties of cloning. There's a strong possibility that the project will misfire as edgy and fun but will add to the confusion and fear at a politically precarious moment.
>
> (19 April 2004, p. D-1)

This is an interesting assessment of *Godsend*'s potential impact on public understanding of cloning 'at a politically precarious moment' in California (see Chapter 1). Nevertheless, most reviews of *Godsend* have regarded its overwrought portrayal of the mad scientist as clichéd and humorous. The narrative revolves around the idea that it may be possible to clone a deceased child, using somatic cell nuclear transfer, providing that cells are harvested within days of the child's death. In our earlier discussions of this film in Chapters 4 and 5 we noted the employment of iconographic microscopic video images associated with contemporary cloning technoscience, as well as its portrayal of feminine infertility desperation. More generally, *Godsend* figures the main scientist, Richard Wells, as bad, mad and possibly supernaturally evil. The film charts an unsteady course between genres, sometimes operating through mimetic realism, sometimes becoming near-future speculative or science fiction, and at other points, transforming into a thriller verging on horror. This genre-blending does not seem to be offering a carefully crafted meta-commentary but, rather, it seems to be the result of a weak script. Many of the reviews rebuke the film for its inept reworking and comment about its narrative and visual cobbling-together of genre clichés. Hence, *Godsend* has been variously referred to as science fiction, thriller and horror, or simply as 'an unresurrectable muddle' (*Austin Film Chronicle*, 30 April 2004).

Somewhat surprisingly, the film's promotional campaign was more imaginative and rather more effectively executed than the screenplay. Prior to the US release of the film in (April) 2004, the distribution company, Lion's Gate, launched a website advertising 'The Godsend Institute'. This title refers to the institution and location in the film where reproductive cloning (via SCNT) is performed. This website for The Godsend Institute had few immediately obvious links back to the film site, or to New Line Cinema or Lion's Gate, and it was presented, and purportedly read by some, as an authentic service offering to clone the children of recently bereaved parents in order to replace their dead children. The site, which was still accessible in March 2007, hosts testimonials by allegedly satisfied customers and explains cloning techniques in some detail, relating these to the cloning of mammals through cell nuclear transfer or cell nuclear replacement techniques. The site appeared under sponsored adverts for fertility and cloning services and still appears in online fertility directories.

Another site set up by the film marketing company (evidenced by the domain registration) was a petition site called 'Stop the Godsend Institute'. This site had no direct links back to any of the explicitly sanctioned film sites, although there was a link back to The Godsend Institute. The creator of the petition purported to have read the Institute site as authentic and protested against its work. The petition claimed to be targeting a US government agency with a view to closing the Institute down. The petition site gathered over 650 signatories.

The Godsend Institute site established plausibility through the use of high production values and the generic techniques of corporate website production. Evidence-based materials such as testimonials from satisfied customers, photographs and contact details also lent it credibility. A significant element was the site's almost complete dissociation from the film merchandising cluster and its explicit distancing from the repeated cloning claims of Clonaid.

There was also a page given over to a brief explanation of somatic cell nuclear replacement techniques. This was a description which draws on recent practices, most notably the techniques which have been successfully employed on non-human animals by Ian Wilmut and others. By 2004 when this site first appeared, the elements of these practices had been established and had been widely circulated within the science community, government agencies and the mass media. By this time, the use of SCNT techniques in relation to non-human animals had become a relatively familiar (although statistically unsuccessful) practice.

Through its knowing manipulation of the genre conventions of health services online marketing and online petitions, this promotional strategy around *Godsend* makes available a range of reading positions that depend on whether the reader approaches the sites as a sophisticated consumer of multimedia intertextuality, a consumer of infertility services or a critic of such services. Read as a genuine marketing address from The Godsend Institute, the website offers a service that draws on discourses already circulating about human reproductive cloning, by seeming to bring the imagined future into the (fictional) present. The key elements of these discourses are that human reproductive cloning is or soon will be available via SCNT, that human reproductive cloning produces a copy or twin of the clone, that bereavement produces a need (loss) that can be met by this technological application and that the needs of this segment of the population (will be) have been/can be successfully met through this application. Taken as genuine, The Godsend Institute website could either attract potential consumers or evoke anger amongst opponents of reproductive cloning.

Alternatively, read as a witty and inventive sales promotion designed to attract media-literate consumers to view the film, the thematic focus shifts. The key elements for engagement are that a film company has set up an inviting set of websites, the sites are constructed through the aesthetic codes of realism, and they are designed to generate controversy and increase the film audience.

The first reading draws on a narrative progression from normative status quo (happy heterosexual family) through disruption (death of the child) back to normative status quo (return to the reconstituted, but aesthetically similar, happy heterosexual family). In effect, this rehearses a version of the 'baby without blemish' story discussed in Chapter 2 (Mulkay 1997: 69–82). The fear which is played on through the provision of this commodified service is the fear of the loss (death) of a child and the associated fear of the destruction of the family. This fear is then retrospectively converted into a need that can be met by human reproductive cloning. Viewers of the site who do not share this predicament are invited to identify with the plight of bereaved parents. Several elements of the scenario are naturalised in this narrative: that children outlive their parents; the constitution of the heterosexual family; and the pursuit of the technological fix.

The second reading does not invite interpellation into a family drama or bioethics but instead involves a more critical view of the media and offers identification with a sophisticated reading position. The questions raised through this are about the effectiveness of such advertising, realism, genre, the ethics of marketing, and media literacy. The petition website made available complementary reading positions, again depending on whether viewers/readers understand it to be a genuine opportunity to engage in online activism or as a promotional activity designed to generate controversy.

Taken together, and in dialogue with the feature film, these websites provide a means for readers/viewers to engage in novel forms of debates about scientific and media practice. They did generate debate in web logs (blogs), news agencies and Australian and US bioethics publications, as well as garnering attention from the film advertising industry. Obviously there are questions about whether these websites were understood to be genuine examples of the genres they mimicked, but here we draw attention to the ways in which they form part of the context for communications about cloning in complex intersections with genres that are more unambiguously identified either as factual (news reporting) or fictional (Hollywood film).

Panos Zavos's websites: a 'maverick cloner' and warrants for truth claims

We have already discussed the ways in which journalists and some of their scientist sources have characterised Panos Zavos's media addresses on cloning as self-publicity rather than science communication. Zavos is an intriguing figure because he has pursued a project of scientific self-legitimation through media relations and other public relations interventions that have, in the UK in particular, resulted in him being portrayed as a fraud and a maverick. Despite this, he mimes a strategy that seems to fit him for many of the criteria for which the UK media – and indeed UK scientists – represent him as unfit. His websites seem to crystallise this media relations strategy rather neatly, so we analysed the versions of his webpages extant in October 2006.[5] In the analysis to follow, we examine the ways in which Zavos appropriates, reworks and responds to this figuring of his reputation on the websites zavos.org, the umbrella site for all the infertility-related services he provides, and reprogen.org, the website for his human reproductive cloning business. On these sites a complex recirculation of mediated communications about cloning is effected in order to represent Zavos's contributions to cloning discourse as credible. As part of this process, specific reference is made to his participation in a range of media genres.

The website www.zavos.org brings together the multiple web presences of Zavos's diverse business interests in infertility treatment. The front page of this website asserts the following claim in its banner headline: 'Being first is what it's all about! Dr Zavos – the man that created the first Human Cloned Embryo'.[6] The page is then divided into three main horizontal sections. The top section offers a welcome to the Zavos organisation, and also the opportunity to click onto another page to view a list of Dr Zavos's achievements. However, much more prominent in this section are links to five publications. The first three links provide a mini-narrative of cloning discourse since Zavos's entry into the field:

- Towards scientific discussion of human reproductive cloning;
- Human reproductive cloning: The time is near;
- Cloning and cheating.

These link to two commentary pieces published in *Reproductive Biomedicine Online* (*RBM Online*); one authored by Joe Leigh Simpson (a medical researcher and practitioner at Baylor College of Medicine, Austin, Texas) the other by Panos Zavos himself. There is also a link to an editorial by Robert Edwards, *RBM Online*'s chief editor, best known for his involvement, together with Robert Steptoe, in the birth of Louise Brown – the first IVF baby. The last link is particularly interesting, because it cites an article by Zavos, as well as one from Miodrag Stojkovic, as reporting on the true state of current cloning practice – 'reporting the early demise of cloned human eggs'. These articles are cited in contrast to the fraudulent claims of Hwang. The various components of this website suggest that it is possible to construct Zavos as less of an outsider to the reputable bioscientific community than his figuring in the UK news media suggested. By linking his own website to the *RBM Online* article, Zavos seems to be seeking to legitimate his claims about cloning through an emphasis on his publications in peer-reviewed journals. The fourth link also underwrites this reputation as it is the full citation reference for a research paper co-authored by Zavos, Karl Illmensee and a third researcher, M. Levanduski, which had appeared in the journal *Fertility and Sterility*.[7]

The second section of the webpage is devoted to links to five realplayer video clips of US television news bulletins of Zavos interviews. The titles under the videos identify the interviewers and the news channels that they represent, but again no date of broadcast is given. This is a crucial omission on a website, as timeliness is a sign of value on a well-maintained website as it is for news journalism. In fact, the interviews posted here were all conducted in January 2003, on the day that Clonaid claimed that the first cloned baby would arrive in the USA from the unspecified country where she had been born. These videos seem to have been used to establish Zavos's claims about cloning as truthful, partly through explicitly positioning his claims in contrast to those of Clonaid. In the interviews Zavos uses many of the same strategies to distinguish himself from other putative human reproductive cloners that UK scientists engaged in fertility and stem cell research (therapeutic cloning) used to distance themselves from him. He castigates the claims of these would-be cloners as fantasies, as unverified and unverifiable. He makes similar assertions about Severino Antinori and explains that he ended his working relationship with the Italian 'fertility maverick' because of the latter's false claims. Zavos contrasts himself with Antinori by saying that he would abide by a scientific culture of evidence as and when he was successful in human cloning. He also contends that Brigitte Boisselier is simply a compelling media spokesperson rather than a genuine scientific researcher. Very similar accusations, of course, have been levelled against Zavos.

The final section on the Zavos.org homepage which performs his credibility as an expert in the cloning market is a series of links to a set of fertility-related businesses in which he has financial interests. These include the Kentucky Centre for Reproductive Medicine and IVF, which provides 'conventional' assisted reproductive technologies, Reprogen, the organisation that offers 'Parenthood through reproductive cloning', and a range of businesses that offer diagnostic devices and tests by mail order. Since IVF research and services have predominantly developed in the commercial sector, particularly in the USA, this section appeals to those visitors who wish to see human reproductive cloning realised as a commodified service.

The front page of reprogen.org, the website for Zavos's human reproductive cloning business, makes the bold declaration: 'First Human Cloned Embryo: The Time Is Near!'. In order to enter the website it is possible to click either 'Click to enter' or 'Press Releases'. Immediately to the left of the 'Press Releases' text there is an animated text announcing that the Press Releases are 'New'. Perhaps this was intended to read 'News', as 'News' is a conventional category on the front page of corporate websites. In fact, the only press release posted is dated 2 February 2004, which was the day Zavos announced that no pregnancy had been established, following his previous claims (17 January 2004) that he had implanted a cloned human embryo. Indeed, the second (February) announcement forms the conclusion of the press release, but the preceding section reports on the January press conference. This would no longer have been newsworthy, prior to the announcement of the failure of the pregnancy, and, in any case, the press release was mounted on the website months later. Nevertheless, representing himself as somebody who can reliably attract journalists to report on his claims is a strategy for representing his work as authorised by arbiters of truth. Zavos operates two more of his key strategies for presenting himself as making credible and responsible scientific claims in this press release. He makes statements inviting government regulation of human reproductive cloning (as an alternative to banning it). He also states that he will publish his work in a peer-reviewed journal.

Overall, however, Zavos's interesting online public relations strategy is badly let down by its production values. The website is poorly designed, with fussy and distracting graphics and crucial words denoting expertise and credibility are misspelt. On the homepage 'achievements' is rendered as 'achivements' and if you follow the link to 'Publications' you find them listed under the title 'Manuscripts in Referred Journals' instead of 'refereed'. As a site, Zavos.org compares unfavourably with godsendinstitute.org. Of course, the latter is a Hollywood film industry production which has probably been produced by highly paid media professionals, whilst it appears that Zavos.org has been designed by a freelance operating without this infrastructural support.[8] Nor is the site for The Godsend Institute subject to the same requirement for constant overhaul as Zavos.org. Zavos points out:

> I have a full-time webmaster that babysits our website. And there are times that we don't have time to put up what I've done the last two or three weeks, for my trips, for instance, my presentations, my honours that I receive. And it's very, very difficult. It can be various people's full-time employment and commitment that I do not have. And don't forget that I have to make a living.
>
> (Interview with authors, March 2007)

Paradoxically, considering the accusations made about Zavos's commitment to self-publicising, it appears as though his credibility might be better served by those media genres which operate independent peer review or editorial control than by a genre over which, theoretically at least, he has full authority. Deploying generic codes and production values that underwrite credibility is a skilled and specialised task with its own demands that are not always reconcilable with the institutional goals of scientists, maverick or not.

Conclusion

In this chapter we have reflected on strategies for making some claims credible, while others are invalidated. We first examined some of the strategies used in mainstream science and in news reporting – specifically focusing on the claims made by Hwang. We then introduced three case studies in what we have called secondary or liminal media, because of the interpretative skills they demand from their audiences. These skills are arguably more specialised than those demanded by the media we have nominated as primary because of their central occupation of the imagined terrain of the mediation of science. We have explored the ways in which different media genres may be used to lend plausibility or authority to the claims of specific cloning scientists both in the rhetorical, narrative and visual tropes they deploy and through the particular values associated with each genre. We have also noted that mediation of cloning claims is taken for granted – in some imagined public space of consensus – as a transparent process of communication until crises occur. Once a fraud is identified (as in the Hwang scandals), then the mediation itself becomes defined as a problem. In the Public Understanding of Science literature, blame in these moments of crisis has typically been attributed to mass-media genres, which are imagined to occupy a lower position in the hierarchy of truth-telling than specialist scientific journals. However, even while identifying the media circus around Hwang as part of the problem, the Hwang scandal also turned the self-reflective gaze of the media onto the specialist scientific journals and their relationships with other genres in the contextual web of science. This raises fundamental questions about how facts and fictions are verified and underwritten by material and cultural practices.

In this chapter we drew back from our focus on key media events in news and Hollywood film. We have attempted to fill in more of the contextual web of *media* in which these privileged genres are located and to indicate our awareness of a complex nexus of inter-textual and extra-textual relationships that constitute the contemporary cloning imaginary. This imaginary is constituted from a complex layering of messages and meanings temporarily fixed and periodically re-figured as the temporal horizon between scientific speculation and scientific realisation shifts. There is no one moment or single privileged site at which we can achieve access – mediated or not – to the 'true facts' of cloning.

In Chapter 2, we explored contemporary meanings of cloning through the juxtaposition of a series of genealogies produced by scientists, journalists and bioethicists, in the process demonstrating that there are multiple versions of the history and emergence of this field of technoscientific endeavour. In Chapter 3, we explored how the future of cloning was in the process of being imagined in multiple sites, including public lectures, film festivals, a commemorative stamp, national policy documents and expert committee proceedings. These events and documents continue to circulate as they are re-mediated through web presences or through the citational practices of successive media productions. In chapters 4 and 5 we focused squarely on the two key sites of meaning production that are evoked when claims are made about the need to inform 'the public' about science, and the necessity of calming fears and dispelling misapprehensions caused by fantastic fictions or ill-informed news stories. (There is more on 'the public' in Chapter 7.)

We believe that our review of the centrality of processes of mediating the Hwang story provides a useful context for interrogating the performative nature of cloning claims in the media. We hope that the three examples of liminal genres that we have discussed in this chapter demonstrate that purification of the boundaries between cloning facts and cloning fictions is unlikely ever to be achieved. The intertextual and cross-generic borrowings are intrinsic to the mediation of science-in-the-making because speculation and contestation have not yet given way to the 'hegemonic currency of science'. In the arena of cloning, it is hard to imagine that either the techniques or the ethics of the field will ever be sufficiently 'black boxed' so that cloning fantasies and respectable cloning science will be hermetically sealed off from each other. That said, the processes of boundary management that we discuss in this book represent the attempts of a whole host of interested parties to move towards fitting a black box around cloning. Nonetheless, the Hwang affair also demonstrates the potential intractability of material objects in moving towards this closure; effective communication is not a panacea.

7 The constitution of publics and audiences

Introduction

The intersections of genomic science and cloning have generated distinctive forms of public interest and concern. As such, they have posed particular challenges to efforts to democratise science and to sustain public engagement with science. Against this backdrop, we explore recent rhetoric about publics pertaining to cloning and genomics, considering in particular the conceptualisation of citizens, consumers, patients and stakeholders.

We examine key issues of agency and structure, ethics and politics by examining how publics have been positioned in recent policy statements, news reporting and documentaries on reproductive and therapeutic cloning. We also examine science-fiction scenarios in film, television drama and drama-documentary. Our analysis incorporates some reflections on focus group and interview material, as we investigate how people, including scientists and 'ordinary people', talk about public understandings of cloning.[1]

In recent discourses of human cloning we argue that publics are figured through media representations, policy documents and audience responses as:

- global humans;
- rational, autonomous, deliberative, national citizens;
- irrational extremists.

The first figuration was discussed in Chapter 3 and we do not return to the construction of the global human subject in great detail here. However, as discussed earlier, the casting of such global subjects is usually integral to the figuring of national citizens and individuals. In the following section, we thus focus on analysing the second and third of these ways in which publics figure in debates on human cloning. First, we analyse the figuring of autonomous, rational, biopolitical citizens (Rose 2001; Rabinow and Rose 2003). Second, we look at the 'privatised' others of this imagined polity – the irrational extremists. We are interested in how it is that some actors are evicted from the political terrain of science, while others are designated as acceptable occupants of this space.[2] Through these processes of exclusion and legitimation, public opinion and objective value are mutually constituted. In the recent casting of publics, the rational autonomous citizen and the irrational extremist may be figured as individuals and/or groups. However it is only the rational, autonomous citizen who appears as a biopolitical subject: it is only this subject that can speak (Spivak 1988; Butler 1997). Those cast as irrational extremists are denied status in such conjurings of the public through a discourse which identifies feminised bias and particular interests as private and which draws on established oppositions of: public with private, public interest with bias, and rationality with emotion.

This contrast can be vividly illustrated with reference to two examples of representations of confrontation: one from a recent film, the other from a recent television drama-documentary. In a scene from the Hollywood film *The 6th Day* (2000), protesters are shown gathering outside the buildings of a biotechnology company and they are represented as noisy and incoherent. The discernible images are of people carrying banners. They are, both audibly and visually, in the background. The faces of the protestors are not distinguishable: they can only be seen as a crowd and as an ineffectual assemblage. The point of identification offered in the film is with the hero/protagonist Adam Gibson who stands in for the rational citizen. In the US context in which the film is set, Gibson appears as the action hero who takes his citizenship into his own hands and fights off the evil of unregulated greed.

In the UK BBC drama-documentary *If . . . Cloning Could Cure Us* (2004), the faces of the protestors are also indiscernible since they either wear masks or hide behind doors. They move rapidly, shouting and throwing objects. They are only made visible through the traces of their actions such as graffiti. They persecute one of the main protagonists – a wheelchair user seeking a stem-cell cure – by spray-painting buildings. Indeed, at one point, they spray red paint onto his body through the letter-box of his home. *If*, in contrast with *The 6th Day*, casts the protesters as more sinister and irrational extremists, as they are portrayed as both visible as a force, but invisible as subjects.

While quantitatively and qualitatively science and governmental policy have dominated recent discourses of cloning, audiences have positioned themselves in normative, negotiated and alternative ways. For this reason, we felt that it would be useful to bring together two rich and diverse but different analytical frameworks: science communications and media studies of audiences. Science communication has offered specific modes of conceptualising the science and society relationship through different models of disciplinary formation, education, advertising, public relations, communication, deliberation and engagement. Audience research has generated distinct models of characterising the audience from the 'mass' to the differentiated, and from passive to multiple, active and creative. We draw on these two fields in this chapter, focusing on a few significant intersections and points of divergence, to investigate how publics have been constituted in recent cloning discourses.

Rational autonomous acquiescence

As discussed in Chapter 6, the drama-documentary *If* used a fictional drama to raise the question: 'What if cloning could cure us?' It did this by developing the story of an embryonic stem-cell scientist trying to cure a patient who had been paralysed in a climbing accident. Reiterating the 'miracle' cure trope through this story, the programme also included a series of interviews with scientists and commentators including the then head of the HFEA, Suzi Leather, and a representative from Comment on Reproductive Ethics (CORE), Josephine Quintavalle. These interviewees were asked to comment about the scenarios played out in the fictional drama as if they had been real.

The dramatic device for representing rational and deliberative public opinion in the drama was the inclusion of the jury, who were asked to decide whether the scientist was guilty or innocent of conducting illegal embryo experiments. This format makes direct reference to the recent UK and wider European policy context in which citizens' juries have become a recognised and familiar method for encouraging public engagement with science. Moreover, this element in the drama had particular resonance, since the jury is generally identified as a cornerstone of democratic justice, characterised traditionally as 'twelve good men and true'. There was also another version of the public represented in this programme in the images of the activists, who were both visually coded as embodying 'mass' hysteria and displayed as marginal. In addition to the on-screen jury, the off-screen television audience was invited to register their vote on the same question.

The jury and the activists were represented visually in contrasting ways. In the case of the jury, the camera played slowly over each individual character. The viewer was offered the opportunity to imagine each jury member as a thinking subject and the collective (of the jury) was thereby individualised and represented as deliberative. The activists, in contrast, were only seen *en masse* or in action, often through peripheral vision. The jury members were positioned as invited participants with their own space, whilst the activists were displayed as disruptive intruders.

Two endings to the programme were produced, although only one was screened, with the phone poll determining (with a verdict of either 'guilty' or 'innocent') which ending would be shown. The audience 'at home' voted 81 per cent (9,381 votes) for 'not guilty' and 19 per cent (2,235 votes) for a guilty verdict, and so the latter version was broadcast. This programme was unusual in that it mixed drama and documentary, fact and fiction, in staging a citizen jury activity as a media event. However, it was not unusual in the way that it offered human cloning as a story about the rights of individuals to the hope of future cures, posed in opposition to the moral outrage of unidentified and virtually unidentifiable, irrational activists. The use of the jury to represent the public and of alternative endings decided by voting illustrates the deployment of the public as arbiters of value or a new kind of objectivity associated with the recent concern for public engagement with science in the

UK and the USA (Barthe and Lindhardt 2006). We discuss this coupling of objectivity and publics later in this chapter.

In news stories about therapeutic cloning, the public is generally represented through the figure of the potential patient. The patient or 'victim of disease' is described as a hopeful sufferer and as, at once, both part of an interested group and a general category (anyone with almost any health problem). The group or category of people which it is imagined will be treated through stem-cell technology has expanded considerably during the past decade. This expansion registers the investment in the potential of the stem cell within technoscience, which, as we have argued, is viewed as almost omnipotent and associated with the trope of regeneration. As the source of both embryonic stem cells and DNA-matched stem-cell lines, therapeutic cloning can be framed as of interest to a huge audience. However, to appeal to this audience, therapeutic cloning must be rendered intelligible as curative and cures must be foregrounded as the ultimate goal.[3] Despite the fact that cures have not yet been realised through this technology, this intelligibility is sustained through the representation of an ever proliferating list of diseases and damages to the body which might be curable through stem-cell technology.

In recent UK policy documents, publics appear as the figures of deliberative citizens who may have concerns about cloning, but who will be open to explanations about why human cloning is necessary. One of the findings of the House of Lords Select Committee on Science and Technology Third Report, *Science and Society* (14 March 2000), was that scientific research would gain more public support if it were linked to medical benefits (*New Scientist*/MORI 5–8 March 1999). In the context of a political culture in the UK which focuses on outcomes, this has been widely taken up in science and policy contexts, to the extent that proposals for research are expected to delineate claims regarding health-care benefits or therapy. Hence, the orientation of address is towards a rational individualised public composed of those who will understand the promise of health benefits and therefore support therapeutic cloning. The construction of this subject position was apparent in UK Prime Minister Tony Blair's address to the Royal Society in 2006, as he reflected on the prospects of public support for stem-cell research:

> Then, we need to engage the public at a very early stage. The reaction to stem-cell research gives us grounds for optimism. Unlike with GM foods, the public were engaged early enough and the argument has duly been conducted rationally. We then need always to be clear about how the benefits accrue to individuals. The anti-GM lobby does not campaign against GM human insulin because the benefits to people with diabetes are obvious. The acceptability of stem-cell research links to the fact that people can see how it would help treat illness.
>
> (Blair 2006)

This subject position has been conjured repeatedly in speeches, commentaries and literature on public engagement, consultation and understanding. In the

UK this conjuring is part of a political culture of relative homogeneity and empirical rationalism around biotechnology (Jasanoff 2005). This contrasts with the political culture of the USA (ibid.), which is relatively heterogeneous, and South Korea, which explicitly focused national support on biotechnology (Helgesen 1995). Such national differences are important and we shall return to them later in this chapter in the discussion of focus group and audience research in different localities. Nevertheless, the appeal to the rational citizen cannot be separated entirely from the construction of a global subject, the interpellation of which is at the centre of Western versions of democracy as a global norm. The global subject and national citizen are, as noted previously, in many cases mutually constitutive as it is in the appeal to the common humanity of the rational citizen that specificity is obscured. The rational citizen and the global subject intersect. In contrast, the constitutive 'other' – the irrational extremist – is never rendered as the global subject and is always conjured as specific and exceptional (Butler 2004).

Publics appear in cloning discourses as individuals who stand in for groups. These metonymic individuals are signalled by the use of terms such as 'each one of us', 'you' and 'individuals'. Likewise, generic, inclusive collective nouns are used to designate groups in such discourse: people, the public, the general public, humans and populations. These generic publics appear as the beneficiaries of science – now and in future generations – as exemplified in the imagined community invoked in the title of the NHS report on genetics published by the UK government in 2003 – *Our Inheritance, Our Future* (see Chapter 3). Hence, publics appear as the recipients of cures in a generalised version of a collective humanity. It is this generalised version of humanity which is invoked in Ian Wilmut and Roger Highfield's dedication of *After Dolly* (2006): 'To the tens of millions of people who will one day benefit from research on cloning, embryos and stem cells.' These metonymic individuals are significant in contemporary discourses of choice because they at once stand for a collective, whilst simultaneously providing the grounds for a reduction of the politics of biotechnology to questions of individual ethics and agency. The public individual is multiply invoked, but singular in appeal. Thus the biopolitics of cloning can simultaneously be an 'ethopolitics' or politics of the self (Rose 2001).

When the rational citizen is represented as belonging to a specific group these are always large, unquantifiable groups such as 'the nation' or 'women'. Thus, as we discussed in Chapter 5, a generalised representation of autonomous agential people is made when women are invited to donate eggs for research. This occurs in the invitation extended by Ian Wilmut, Alison Murdoch and the HFEA in the UK which completely obscured the specificity of the group who are actually being asked to donate – women. This use of the generalised term 'people' obscures those already in the IVF process in clinics in the north east of England who, for a variety of reasons, but potentially because they have had all the cycles of treatment they can afford, may give some of their eggs to cloning research in exchange for subsidised IVF treatment.

However, this linguistic interpellation, whilst obscuring this specificity, also identifies women as the general subjects of cloning discourses, proffering a more flexible future subject position of so-called 'non-patient donors'.

The representation of the public as the nation is a dominant and familiar way of invoking the public as, at once specific, but also, as unknowable (Anderson 1983; Bhabha 1990). Although some have argued that the political power of the nation has declined comparatively in the early twenty-first century, the imagined community of the nation state remains a powerful rhetorical reference point. In recent UK policy speeches, statements and documents, for example, 'we', 'us' and 'our' are used to elicit identification with the nation, as in the series of lectures by Tony Blair entitled, 'Our Nation's Future' (2006).

Alongside these representations of the public epitomised as individuals (or sets of individuals) or as an extensive collective, there have also been representations of specific groups which make up the public. These are cast as sub-groups of the larger public. These sub-groups may stand (in) for the public (similarly to the metonymic individual discussed above) and/or they may act as models for the public. However, they are also understood as a sub-divided, constituent part of the public. These sub-groups predominantly appear in the two different ways already outlined. First, they appear as deliberative citizens such as jurors, voters, local policy makers and/or specific members of interest(ed) groups. Second, they appear as irrational extremists, represented as activists and/or members of interest(ed) groups.

Interest groups and women may be constituted as either rational citizens or irrational extremists, depending on the context. When publics are represented as patient groups, they appear as groups with a special perspective which merits the attention of the wider public. They merit such attention because they are associated with suffering and this may either be made explicit, or it may remain implicit. In the UK, animal rights activists or religious groups (this latter category has a different meaning in the USA)[4] are generally consigned to the irrational category and are, hence, construed as dangerous and as having a negative influence on the wider public. Thus, there have been interesting distinctions at play in the discursive representation of the public as a general group in comparison with the representation of specific groups of the public.

More generally, the rational citizen is constituted as the beneficiary of the technoscientific developments around human cloning. We found that in private, one-to-one interviews, scientists often offered comparatively limited claims about the prospects for cures through SCNT and stem-cell research. Nevertheless, they maintained that SCNT will contribute significantly to the sum of human knowledge and, hence, that cloning is beneficial in an abstract way. In these kinds of narratives the public is invoked as benefiting from genomic technoscience in a mode that is resonant with widely circulated notions of the enlightenment of the global subject through science. This sum-of-knowledge argument relies on a vision of the collective good which is commensurate with a politics of the self. The oscillation between appeals to individuals and appeals to amorphous collectives invokes a vision of the collective which simply scales-up the notion of the autonomous citizen.

Duped resistance

The actions of those construed as activists and/or religious groups, in the UK, are frequently positioned as against the public interest and as acting on behalf of private interest. Activists appear in television drama and film as loud and, often, violent groups who operate through direct action, as is seen in *If* and *The 6th Day*. In these portrayals, direct action is always represented as destructive, unthinking, cruel and often immoral. The representation of activists as dupes or tools who are subject to the manipulation of more powerful forces is also a relatively common trope. In the film *Aeon Flux* (2005) the eponymous main character is an activist working to undermine a corrupt and totalitarian regime. It is through her actions that the human cloning procedures are discovered and eventually stopped. However, her role as an activist in a collective that might have political agency is undermined by the manipulation of the activist group by members of the governmental regime. It is only when she breaks away and operates on her own that this figure emerges as a character who contributes to the development of the plot.

The largely negative figuration of collective activism or religion means that those identified with such activities are effectively expelled from the castings of the legitimate public. In the UK this pattern emerged in the press coverage of the 1990s embryo debates in which MPs and scientists in the PROGRESS lobby helped to secure hegemonic versions of secular rationalism (Mulkay 1997). In this figuration, those not identified with such rationalism are portrayed variously, with some publics represented as having 'private' interests, as activists, religious groups, feminists or 'powerful lobbies' who unreasonably resist the respectable scientific practices of technoscience – in this case, cloning. Hence, they are positioned as *specific* publics (as activists) who are operating against the imagined public interest because of their illegitimate private interests. These groups are represented as *specific* minorities, but the imagery through which they are portrayed is fixed and homogeneous, involving a limited repertoire of stereotypical representations (for example, see *If*, *The 6th Day*, *Blueprint*).

Those linked to particular interests are thus configured as subject to bias and undue influence when they are identified with political activism, religion or feminism in the UK. However, the affiliations (and interests) of investors, scientists, politicians and many patient groups are generally cast as *in* the public interest. Claims to be serving the public interest can thus be powerful but problematic discursive trump cards in techoscientific and health controversies and policy development.

Thus far, activists have rarely figured in UK press and broadcast news coverage of human cloning stories. Although there is, by now, a repertoire of images of activists associated with protests either against GM crops or for animal rights, there are very few instances in which activism around human cloning has been represented visually. The rare instance in which this occurred was in the coverage of the South Korean cloning fraud, which included images showing demonstrations of support for Hwang; for example, queues of women waiting to volunteer for egg donation. However, in UK coverage of therapeutic cloning, there has been very little visual reference to activism, although there have been references to groups cast as antagonistic, such as CORE and Hands Off Our Ovaries.

In the reporting on cloning and stem cells in television and newspapers, the science community has been dominant (Williams et al. 2003; Holliman 2004). When groups or their representatives are seen as posing difficulties or opposition to science, they have generally been acknowledged as voices – often lone voices – but such voices are usually positioned as distant from the legitimate public. For example, in recent news stories about therapeutic cloning in the UK, Josephine Quintavalle is often positioned as 'anti-science' (despite being scientifically well informed). Quintavalle and her organisation, Comment on Reproductive Ethics (CORE), have frequently occupied and embodied this position of the lone voice. In addition, occasionally other groups (homosexuals, women, patients, and religious and ethnic groups) have been represented through the voice of a single figure. Such voices have generally been cast as marginal.

Where activism has been represented as a threat to biotechnology, activists appear as dangerous, although ultimately containable. Specific forms of anti-science activism have come to be regarded by some commentators in the UK as a serious threat. In the UK in the early twenty-first century, animal rights activists are frequently portrayed as posing a danger to the nation, science and individuals. This has often been couched in terms of a danger that must be tackled:

> If we hadn't taken on the animal rights extremists, we might well have lost essential scientific research to Britain with incalculable economic damage to the country to say nothing of the value of the research in the treatment of disease.
>
> (Blair 2006)

Blair evokes the spectre of animal rights 'extremists' as an enemy which is presented, on the one hand, as defeated (having been 'taken on') but, on the other hand, as posing a recurring threat to the development of human biotechnology in the UK. Indeed, animal rights activism becomes the prime reference point as Blair forges a governmental strategy towards science and publics in his 'Our Nation's Future' lecture on science:

> There are two things in particular that threaten the strong position we have attained. The first is perhaps the most difficult issue of all. Government must show leadership and courage in standing up for science

> and rejecting an irrational public debate around it. . . . Yet in many instances, a powerful and vocal lobby, with access to all the media channels and an interest in polarising the argument, frames the debate.
>
> (Blair 2006)

In this quotation Blair conjures fears of what he terms 'irrational public debate'. This allows for only a particular kind of voice in the public arena. Blair's comments here seem to illustrate arguments which Nancy Fraser (Fraser 1992) makes in relation to concepts of the public sphere. She notes how it may be invoked in exclusionary modes but framed in the name of inclusion. Fraser contends that, in stratified societies, facilitating the proliferation of subaltern positions might be preferable and, indeed, more democratic, than simply reinforcing hegemonic normative political positions (ibid.).

Blair's identification of acceptable and unacceptable publics is accompanied by his assessment that the media are part of the problem. Indeed, blaming the media has become an endemic feature of recent cloning discourses. In this particular instance, the media are represented as tools which are under the control of 'a powerful and vocal lobby' which frames the debate. Here and in other instances, when the media and the public are linked, they are represented as problems to be solved. In this discursive coupling, the public appears as the duped audience of the powerful. Hence, activists and the media are cast as passive, as objects or tools which can be (and have been) mobilised by a powerful (but unidentified) anti-science faction. Ironically, such discursive conjuring does not register recent media support for stem cell research.

Discussing public understandings of cloning

In the discussion above we have examined policy making and media texts. In this part of the chapter we consider how people construct publics outside of these specific settings. The following section draws on focus group discussions conducted in Cardiff and London (UK) and Wisconsin (USA), and on individual written responses solicited through the Mass Observation Archive at the University of Sussex, UK (see Chapter 1). In the UK, focus group participants were asked how and what they knew about stem cells and they were also invited to put together a news story about them. These focus groups were given visual images from news reporting to provoke discussion. In the USA the focus group participants watched an episode of the UK television drama *The 11th Hour*[5] together prior to a group discussion about cloning and the media. The respondents to the Mass Observation Archive directive were asked to write about how they knew about 'genes, genetics and cloning' and they were prompted to consider as wide an array of sources as possible.[6]

Like other researchers looking at audiences and publics in this area (Durant et al. 1996; Petersen 2002; Reid 2004; Bates 2005), we found that those who participated in our research claimed that their information about genes, genetics, stem cells and cloning came from two different sources. On the one hand, our participants talked about gaining knowledge of this technoscientific field through personal experience, and on the other hand, they talked about the media as their main source. The media was cited more frequently than personal experience.

The personal narratives which emerged included stories about being subject to testing or illness either directly, or through family and friends. They also drew on experiences in education and employment. For example, one of our research participants was an electrician who had been involved in wiring a laboratory complex in which animal research occurred. This individual explained that he had learned about scientists and their working environment and that he had identified with them, because, as he saw it, they 'were just trying to do their job'. Another worked in an administrative capacity with scientists and commented: 'I completely trust scientists. I consider all scientists to be much more intelligent than me therefore I will follow them blindly.' One focus group participant revealed a complex and multi-layered experience of science through a variety of work experiences. He had been employed as a security guard at a refuse disposal site until it had been closed down when an activist campaign raised complaints about the levels of toxicity at the site. He stated: 'I was doing the security up there like and there were some right head bangers coming [demonstrating], but they shut it down so they had their point proven' (Kegan, Group 14). He explained that this had dented his faith in official risk assessors and made him believe that activists may raise reasonable suspicions. However, since losing his job at the disposal site, he had become largely dependent for his income on working as a drugs trials subject. He observed that this was a position which meant that he had to trust the scientists absolutely, as he was putting his life in their hands. He indicated that his feelings about the promises scientists might offer were further complicated by the fact that he was also a full-time carer for a relative with chronic arthritis.

Other contributors recounted their own science training or the workplace experiences of their family. Some also talked about their encounters with the media industries and, in some cases, with science reporting specifically. One woman described how in a past pregnancy she had a false positive test result for foetal spina bifida. She had continued with the pregnancy anyway and she had only discovered that it was a false positive by carrying to term. Because of this experience she was relatively suspicious of scientific testing. In the focus groups, a number of women told stories about how they had experienced specific medical procedures, particularly those involving testing and surveillance technologies such as ultrasound, with both positive and negative outcomes.

The majority of our research participants cited the media in discussing or writing about their knowledge of human genomics. They discussed genes, genetics and cloning, or stem cells, in relation to representations in both fictional and factual genres. The TV news, the internet, radio, television drama, science-fiction film and novels all figured significantly, and cartoons and adverts were also noted (see Appendix 2). Art exhibitions and science museums were rarely mentioned, although one Mass Observation participant mentioned the artist Hieronymus Bosch in relation to these areas. One focus group member in the USA had attended the 'Gene(sis): Contemporary Art Explores Human Genomics' exhibition. This event had been held in Seattle in 1999 (although it then travelled across the USA) and was an early precursor of the '50 years of DNA' exhibitions held in the UK in 2003.

Rational publics

Audience readings and relationships to recent cloning discourses are varied and a range of motives and experiences contribute to the generation of different reception strategies (Holliman 2004). However, many respondents endorsed the dominant discursive construction of therapeutic cloning as positive. They made comments such as: 'Well, obviously if it saves people's lives, yes, it's good' (Tilly, Group 12) or 'If one single person in this world of ours becomes cured or has an easier life, I think that's a benefit, it's worth doing'. They often positioned stem-cell research in a long history of medical advances: 'Well there's so many treatments that we have now. If it saves a human's life I've got no problems with it really' (Bill, Group 12). Most participants seemed to consider that science required regulation, but most were confident that there were structures in place to ensure this. However, scepticism was often expressed about how these would operate internationally. For most of our research participants, reproductive cloning was clearly viewed negatively and activists and science fiction were often cited as contributing to a misleadingly negative image of science.

Despite this general pattern which endorsed and echoed the dominant discourses on cloning which we have highlighted, there were some rather notably different positionings elicited. Some of those we consulted positioned themselves almost as 'meta-commentators', appraising the public and situating themselves outside or above the public. A respondent to the Mass Observation consultation wrote:

> Cloning full humans is a long way off, and even if many films and books and TV shows are a bit lacking in science, trivialising the issues, nevertheless it gets the subject out there into people's lives. Issues regarding the desirability of banning, or regulating, or encouraging research and eventual implementation need to be the subject of public debate.
>
> (MO M1201)

Similar ideas played out in some of the focus group discussions. One participant, 'Miriam', stated that 'You couldn't leave it all to the layman to make the decision because they don't know enough about the implications'.

However, others in the group suggested that the issue should be subject to a public referendum, accompanied by public education:

> Kegan: I think they should have the public vote and they should provide the information with a public vote. . . .
> Miles: But then the people should have the knowledge. . . .
> Miriam: And the thing is you've got to think about the general public, how many people – if that information was put through your door, how many people would actually read it? . . . And how many people then would understand it?
> Miles: There'd have to be programmes on the telly so you could watch it, so you could make –
> Kegan: Like *Eastenders* and *Corrie* [popular UK TV soaps].
>
> (Group 14)

Responses such as these identify the public as 'out there', as something of which their speakers are not a part, and media forms were often cited as appropriate vehicles for getting decision making about genomics and genetics 'out there into people's lives'. Raising the topic for discussion is seen positively and eliciting public participation is assumed to be important. In many respects, such comments seem to echo Tony Blair's invocation that science needs to be taken out there – to 'the street' (2006). However, a significant difference is that, according to Blair's comments and recent UK policy and scientific discourses, advocating the desirability of banning research is not an admissible position, although temporary moratoria may be appropriate.

In the Mass Observation commentary quoted above, the implied subject position of the respondent is not 'out there' in or with the public. Rather, as the overseeing 'meta-commentator', the respondent is adopting the preferred subject position offered in established cloning discourses which imagines a few rational experts, a general public that may be in the process of becoming rational through engagement (seen in this case as being facilitated through media dissemination and debate), and the irrational minority. Through meta-positioning this commentator identifies and aligns with the expert group through the capacity to present this conventional, hierarchical configuring. Other respondents used similar discursive techniques when they deployed the language of observation and research to distance themselves from 'the general public', in comments such as the following: 'The problem is whether the "man on the Clapham omnibus" is in a position to debate the issue. Does the general public really understand the science? – probably not.' (MO S3375).

In the focus groups there was extensive discussion and disagreement about the extent to which the public was in a position to 'debate the issue'. There was broad agreement that participation and engagement were positive, and our research participants reported that they felt able to discuss a variety of opinions. However, when prompted to respond to the question of how science should be regulated, ambiguity as to the value of public opinion was expressed.

At this point in one of the focus groups, there was a move to defer decision making to 'experts' who could take the time to understand the issues at hand. While a deferment to expert decision making came in when discussing regulation in the focus groups, there was almost a consensus amongst those we consulted that the prospect of health benefits made most research legitimate. Hence, it was generally considered that it was worth 'taking the risk' ('bad always comes along with good') with human genomic technoscience. There were significant exceptions to this position that we explore below. However, the conclusion drawn in UK science and policy contexts, that cures sell science (House of Lords Select Committee on Science and Technology Third Report, *Science and Society* (14 March 2000)), appears to be consistently borne out in these responses. This is despite concerns expressed by Robert Winston (2005) (which we present below), that managing expectations in this area is crucial.

Some participants articulated cynical positionings. They could be described as reading against the grain, or, in Stuart Hall's terms, as taking negotiated or oppositional positionings (Hall 1997). Such commentators are effectively negotiating alternative rational subject positions, as is evident in this excerpt from our Mass Observation directive archive:

> One wonders how much completely pointless work is carried out in the name of science just to satisfy a scientist's curiosity . . . their attitude seems to be that unless we allow them to experiment along certain lines [for example, cloning and gene manipulation] then we will be deliberately forfeiting our health in the future.
>
> (MO H3459)

This response could be characterised as simultaneously engaged, rational and opposed to 'cloning' and 'gene-manipulation'. In relation to recent dominant discourses of cloning, this is a reading against the grain because it challenges the conventional perceptions of scientific innovation and the orientation of much recent UK science policy. Yet the image of endlessly curious scientists and of their 'groundbreaking' search 'to push the boundaries' was often viewed in this way. For example, a recurring concern expressed in the focus groups was the suspicion that scientists might not abide by limits imposed on them. As one participant declared: 'Scientists don't just stop there, do they? They've got to take it one step further' (Mary, Group 13). Scientists' creativity and passion was seen as potentially dangerous and as leading inevitably to the risk that someone, somewhere, would try reproductive cloning. Another participant hypothesised that this would happen, conjuring how scientists might respond and have responded to regulatory restrictions:

> Bill: If I was a scientist and somebody said to me – if somebody said to Louis Pasteur: 'Actually it's unethical for you to look at this', he wouldn't have stopped. . . . if you're really excited and you've got the buzz and you want to do something, you're not going to suddenly say

> 'Actually it's unethical, I'll stop doing it.' You might say 'Actually it's unethical, I'll stop doing it out front, but maybe I'll do a little bit in my lunch hour.' But because you're that . . . you can't just turn off, you're either creative or excited . . . if you're really excited or passionate, you're going to see it through whatever really.
>
> Harry: You can't actually turn off science . . . I think people are probably tinkering with it (reproductive cloning) now.
>
> (Group 12)

It is interesting to note that the image invoked here is close to the one that scientists themselves and others sometimes use in engagement activities designed to interest children in science. When science is pitched to children in engagement activities, imagery indicating 'a scientist's curiosity' is often used and curiosity is treated as a positive attribute. Images such as that of a light bulb going off in the mind are used to convey the fun and adventure of science. However, as the quotation above suggests, such curiosity can be reinterpreted as contributing to a lack of restraint and to a resistance to ethical regulation.

The position articulated above resonates with Robert Winston's (2005) warning in an address to the British Association for the Advancement of Science that the scientific community must manage expectations about human genomic research and that the public may become cynical if claims about cures are 'hyped'. The respondent quoted above can be seen as attempting to construct an expert position by claiming reading competencies in relation to the mediation of cloning. The comment cited indicates a refusal to be duped and a refusal to take matters at face value. Indeed, the commentator reads cynical motives into the attempts, by scientists, to claim that cell nuclear transfer experiments will save lives. Implicitly this positioning calls the bluff of those who, according to the recent dominant discourse on cloning, declare that lack of support now will endanger lives in the future.

Some other participants effectively opted out of recent discourses of cloning through their assertions that cloning was either fictional, impossible, to be realised only in the distant future, or through declaring that they found it irrelevant, inaccessible or uninteresting. On occasion, respondents provided alternative agendas or hierarchies of relevance. One respondent insisted that there were far more pressing social issues to be addressed, citing poverty and war, while another group of research participants wondered whether stem-cell cures might, in any case, only ever be available to the wealthy.[7]

Irrational publics

While references to irrational publics were prominent in the cloning discourses of media and policy makers as shown above, they were also prominent in the comments of our research participants, although they were manifested in diverse ways. For example, one of the Mass Observation respondents used the term 'terrorists' to designate animal rights activists. The rational 'meta-commentary' public position was taken up, as this commentator complained of: 'too much giving in to animal rights' (MO A3434). In the UK focus group discussions, animal rights activists were frequently cast as terrorists and irrational, as were direct action anti-abortionists, in some cases. In one group, for example, participants criticised the 'illogical' lobbying groups (referring to animal rights and anti-abortion activists). They highlighted the need for the government to make objective information about the facts available to citizens, as this excerpt indicates:

> Theo: I think a lot of lobbying groups – these groups against abortion and animal rights . . . Well, animal – I have great things with animal rights turning up at the airport. We've had a nightmare with them at the airport [where I work]. They've got this thing with animals going abroad, but they [the protestors] can't understand that [it is] people's pets are going abroad, they just don't like to see dogs and cats flying.
>
> Interviewer: And how do you link that with the sort of stem cell debate?
>
> Theo: Because you're going to get groups who are anti-[stem-cell research] lobbying. [But what you need is to do is] – bring leaflets out, say 'right this is what stem cell is', and put it in local papers.
>
> Mary: To educate people then really I suppose.
>
> (Group 13)

A similar discussion occurred in another group, with unreasonable activists identified as blocking rational debate and progress:

> Harry: Some people get so upset and so agitated about a bunny having perfume put in its eyes that they would actually kill a human being, and they do not see the dichotomy of what they're doing. They'll actually plant bombs and send bombs to scientists.
>
> Imogen: [It's like] the anti-abortionists, isn't it?
>
> Harry: Yes, they will shoot doctors. But I've always found that quite fascinating that someone can say: I'm going to save the bunnies. I'm going to send a bomb to the humble scientist which might kill his five-year-old kid . . . if the scientists are right and they can use these stem cells to repair people and cure disease, then that's obviously a good thing. But . . . maybe it's something we're not ready for and maybe, again, like animal protesting and like abortionists, if they do start doing it, you can bet your bottom dollar you'll see scientists being shot left, right and centre and idiots posting bombs.
>
> (Group 12)

The image of the demonstrators, as represented in the *If* docudrama as faceless, violent and uncaring fanatics, was explicitly employed in this context. In some parallel research to our own, Grace Reid ran focus group discussions after showing a video of the *If* drama-documentary. Her research provides similar insights. In one of the groups, for example, the representation of the demonstrators in this drama-documentary was explicitly identified as realistic and these figures were contrasted with people who support stem-cell research and 'that actually care'. The following excerpt indicates this perception and framing:

> T: I think it [referring to the programme, *If*] just showed, I think that's how the issue in real life sometimes comes across, whereby that fanatical people tend to be from that side of the spectrum. So I think they were just trying to show that, that if this was a real case, people in the courtroom would be throwing stuff like that and would be calling them murderer and would be upset. The people who . . . are on the paraplegic guy's side would then be more subtle, just like the lady who was supporting the research.
>
> E: I think that's a very good point; that you do get the fanatics and then you get the subtle people that actually care.
>
> (Focus Group 1, Reid 2007)

However, this sort of representation of the anti-stem-cell position was actively rejected by some of the Catholic participants in Reid's study. They complained that the voice of opposition to stem-cell research was misrepresented in this programme. They commented, for example, on ways in which 'pro-life' views are 'invariably trivialised because they are portrayed as the views of a minority group and therefore *not* worth listening to'. Reid's Catholic research participants particularly objected to the representation of the demonstrators:

> K: The only people who were being unreasonable were the extremists who were against it . . . they were like the animal liberationists. They were chucking bombs, not bombs, but they were chucking blood. . . . There are people who have just as extreme point of view from the other [pro-stem-cell research] side but *they* were not put forward as being unreasonable.
>
> (Focus Group 3, Reid 2007)

These research participants identified as a minority group and felt they were framed as marginal or irrational. As one research participant commented: 'In the United Kingdom Catholics are a historically badly treated minority and they don't have a say' (Focus Group 4). Indeed, several of these Catholic research participants problematised the idea of any simple 'scientific democracy' on grounds that this could amount to a 'dictatorship' by the majority. One of them reflected:

> And I think that's why it's important to keep these debates going because in the absence of, well let's call it 'our side' . . . putting our case forward what you'll just end up with is a dictatorship of the others. You'll just get whatever suits the majority.
>
> (Focus Group 4, Reid 2007)

Voting and the collection of 'opinions' was, from this perspective, seen as problematic:

> There's a problem with this kind of voting. Opinions are, to quote one of my colleagues, 'Opinions are like assholes, everybody's got one', and what that means really [is opinions don't] matter in the long view. Looking at the greater view of things: What do we think of death? What do we think of suffering? What do we think of, is there a life beyond this?
>
> (Focus Group 2, Reid 2007)

Similar reservations about popular opinion were raised in one of the focus groups we conducted with Muslim women. After lengthy discussion of the pros and cons of human cloning and stem-cell research, they concluded with comments such as 'But I am not an Islamic scholar, I am not qualified to decide' and 'I would actually give [that question] to somebody who's got a lot of knowledge . . . [like] a mufti. I would actually ask his opinion before I really give my conclusion on that' (Focus Group 17). It was only through consulting religious texts and scholars that they felt the debate could be placed in a proper context and related to fundamental questions and values. Like the Catholics quoted above, however, they suspected that a Muslim view on human cloning would be sidelined in any discussion of UK policy and 'public opinion'.

Different types of discussion emerged within the focus groups we conducted in the USA. In the USA, organised Christian religion and, more specifically, the anti-abortion lobby, occupy quite a different position, in comparison to their situation in the UK. In the focus groups we conducted in the USA, animal rights activists were not mentioned and the anti-abortion lobby was attributed with a significant role in a civil society. Christians were also less likely to see themselves as a marginalised minority (compared to Muslims or Catholics in the UK), but the power of religious groups loomed as much more relevant in discussions about the governance of biomedical technoscience. Hence, there are several significant differences between the figuring of publics in the USA and UK, one of which revolves around religion and/or the 'right-to-life', anti-abortion position.

Religious affiliations are generally perceived as marginal to the debates about science and society in the UK. Although there were some references to religious views, and references to the Pope's Easter Address[8] and 'the church' were made, a secular framing was dominant amongst our UK research participants (except where research participants were specifically brought together for our research through their shared religious convictions). However, in the USA, Christian religious beliefs are highly significant in debates about human biotechnology. In the recent Bush administrations (2000–8). Christianity has been firmly endorsed and, in this sense, it has not been regarded as a minority threat to secular rationalism. Instead, it tends to be considered an integral part of the national culture. These differences were evident in the comments our US participants offered after viewing the UK television drama *The 11th Hour*. In the US groups (in contrast to those in the UK), there was no mention of activists but these participants suggested that certain images – such those of aborted foetuses and naked men – would not have been shown in such a casually graphic manner because of the political power of religious groups in their area in the USA.

In the UK focus groups and in the materials from the Mass Observation Archive, participants often referenced science fiction to articulate what cloning was not. In these references science fiction was often conflated with science-fiction *film*. This deployment of science fiction is similar, as we noted previously, to the way some scientists and some policy makers have recently used science fiction. Some of our research participants used references to science fiction in this way to demarcate the boundaries of acceptable forms of cloning and genomic science and to position other people (and occasionally themselves) as misled dupes of fantasy (mis)representations. The complaint of these diverse commentators is the same: science fiction might mislead the public. Whilst some of our Mass Observation and focus group members referred to fears articulated through science fiction, and speculated about the possibility of such fiction becoming fact, very few of them attributed their own knowledge about cloning sciences to fictional sources.[9] Nor did they often justify their fears about science by reference to science-fiction scenarios. In fact, when they presented fears that they felt unsure about or that could be read by others as overblown – 'stupid fears' or 'paranoia' – they did so apologetically, labelling them as 'like a film'. Thus, they did sometimes, rather ruefully, attribute their opinions to fictional sources. The following comment from a respondent who was entirely supportive of genomic sciences, but who articulated her own fears, illustrates the pattern of such disavowals:

> I am woefully ignorant – I blame my expensive and elitist girls' public school education where domestic science was still the only science taught even in the 1980s . . . as general principles I think it is important to avoid the abuse of the vulnerable; to prevent the work of Mengele or Frankenstein-type mad men; and not to create a uniform designer creature. Obviously these principles are rather over informed by Nazi-induced paranoia and sci-fi.
>
> (MO A3434)

Rather than seeing science-fiction films as *legitimising* their fears or anxieties about developments in genomic science, participants in our research tended to refer to factual sources and draw examples from history or recent news events.

Thus, there were frequent references to Nazi eugenics, the Iraq war, the nuclear arms race, or recent medical mishaps to contextualise their unease about aspects of recent technoscientific developments.

However, science-fiction fans in the USA who contributed to discussions after watching *The 11th Hour* and *Teknolust* identified science fiction, across multiple genres, as a source of scientific knowledge. They did not make specific knowledge claims but they judged the television drama and the film in relation to varying perceptions of scientific accuracy. Participants in these groups also positioned themselves as having greater scientific knowledge than the 'general public' because of their familiarity with genre fiction, including the more traditional forms of the science-fiction novel and short story, and they anticipated that they would be less likely to be misled by media representations than non-science-fiction fans. Participants in these groups also drew connections between science-fiction fans, science-fiction writers and scientists, and they emphasised that in their experience there was much overlap between these groups in that fans are also writers and vice versa, and that scientists also write and read science fiction.[10]

Responses about who publics are, came together with questions about the media and about science in our research on audiences and publics. Many respondents implicitly identified themselves as part of the public by framing their response as unqualified, non-expert or ignorant opinion, as is evident in the comment above about the ignorance of someone who went to an 'elitist girls' school'. Other respondents interjected phases such as: 'like me' to indicate their identification as part of the public. 'How can members of the public know which way to turn? I suspect a lot of us, *like me*, look to our gut instincts and rely on whether something "feels right" or not', one respondent reflected (MO N3181). 'It [cloning] just gives me goose bumps' commented a second (Mary, Group 13). A third announced that reproductive cloning: 'just absolutely freaks me out'. She explained that: 'I can't articulate it but there's something . . . just not right about that, it's wrong' (Miriam, Group 14).

In these comments 'gut' feelings are related to judgements on whether something 'feels' right. Reflections such as these are complex, as the commentators are evidently trying to negotiate legitimate subject positions, invoking feelings and moral judgements to explain what could be regarded as irrational stances.

Some of the research participants identified explicitly as part of the 'irrational public' and they negotiated their reading of cloning rather reflexively with an awareness of how they could be cast. In these instances, confessional discursive strategies were often apparent as participants negotiated their positions within dominant discourses of cloning. This move bears some similarity to the claiming of a position of non-expertise, ignorance or going on 'gut feeling' used in more ambivalent responses. For example, one participant presented herself as troubled by the focus on cures within dominant cloning discourses. She interjected qualifying statements about her position: 'I know this is awful to say' and 'I've never mentioned this to anyone else.'

Nevertheless, she contended that generally people should be left to die of diseases. While these discursive qualifications indicate personal reflexivity about controversial views on life and death, they also indicate the power of dominant discourses on cloning, particularly in relation to cures. They also illustrate the difficulties of negotiating alternative positions in relation to dominant cloning discourses.

Only a minority of our participants distanced themselves from the dominant expectation that cloning and genomic technoscience would deliver cures imminently. In addition to the confessional strategy noted above, nature was sometimes used as an authorising trope when such views were espoused:

> You should let nature take its course really, you know. I think everything happens for a reason and if you start messing around, then one day you're going to create them quite bad, you know.
>
> (Martin, Group 11)

> Human cloning might be the next breakthrough. Why do we need it? Using cells from the clone to cure our illnesses? We all have to die. The world cannot sustain an ever-increasing population where babies are born and nobody dies. It's not natural.
>
> (MO M3640)

This second comment came from a respondent who used DNA databases in police work and thought that these tools were of huge benefit. However, whilst this commentator welcomed genetics in crime detection, she invoked visions of 'the natural' to disavow its usefulness for medical purposes. [11]

The image of the irrational public was both drawn on and renegotiated in these audience discussions. Feelings and notions of the natural were also invoked in explanations of what could be considered to be irrational forms of argument. Direct opposition to dominant cloning discourses could be framed cynically, defensively or be qualified in various ways, but such opposition was often also powerfully argued. Indeed, evidence was used to undermine the cures narratives: 'People should be made aware of the lack of cures' (MO R3198). In these positionings, the management of expectations was an issue and dominant cloning cures narratives were resisted in a number of ways.

Publics and audiences: issues in audience research and public engagement

We argue in this book that mediation is a crucial dimension of the making of cloning science. In this chapter we have looked beyond texts and sites of production to consider sites of consumption in which the meanings of cloning are also constructed. Whilst the first part of the chapter looked at how publics are constituted in key texts, the second part drew on materials from our research participants to develop insights relating to consumption. We looked at what selected audiences said about cloning and at their own conceptualisations of publics in relation to cloning discourses. As these analyses indicate, we regard the conceptualisation of the public as a series of audiences in diverse contexts as a fruitful tool in analysing the making of cloning and genomic technoscience more generally. In this section we examine the results of our analysis in the context of current debates and activities in relation to both public engagement and media audiences.

There is a long and well-established tradition of imagining media audiences as the faceless mass and the irrational crowd that can be transformed into rational agents through a progressive media address. This view still has considerable purchase. As such, mass audiences continue to be imagined and addressed as requiring education. There have been sustained critiques of these versions of the public sphere which highlight the exclusions and normative judgements inherent in such conceptualisations, as well as the erasures of identities they affect. However, this vision has influenced some public understanding of science and public engagement with science research, policy and debate. Indeed, this imagining of the public as mass complements and sustains the familiar tropes of public ignorance of science, the deficit in lay comprehension of science, and the transmission model of scientific knowledge which sees it emanating out from the scientific community. These tropes have been both widely circulated and sharply criticised (Irwin and Wynne 1996; Wynne 1996).

In contrast with these frameworks of a mass audience, recent approaches to the media tend to acknowledge that audiences are active in consumption processes and many link reception to production and content through different models of the circuit of culture (Hall 1997), the circuit of mass communication (Miller et al. 1998) and active audience and fan frameworks (Jenkins 1992; McRobbie 1998; Hills 2002). However, such approaches vary in terms of their attribution of agency either to individual readers/audience members (Hills 2002) or to the structures of the media industries (McRobbie 1998). Such research has influenced our investigations and the analyses undertaken in this volume. Mindful of this rich field of research, we would suggest that the media might be conceived as providing, in Elspeth Probyn's terms, a 'contact zone' (Probyn 2005) in which cloning has been and continues to be constituted. The contours of this 'contact zone' have been shaped and re-shaped. We have identified throughout this book some of the most important processes contributing to this shaping. These include the mediation of IVF (Crowe 1990), Dolly (Franklin 1999; Holliman 2004; Franklin, 1997), and the Human Genome Project (Nerlich et al. 2002), the representation of developments in stem-cell research (Parry 2003; Williams et al. 2003) and human therapeutic and reproductive cloning (Bates 2005; Kitzinger and Williams 2005; Haran 2007). Cloning is also embroiled more broadly, in ideas about disability, medicine and progress. Audiences are active in their readings of human cloning, and diverse understandings and strategies are produced in response to cloning. Nevertheless, these responses are also contingent on the dominant discursive frames established over time and across a multiplicity of texts (Henderson and Kitzinger 1999; Williams et al. 2003).

In bringing together the fields of media studies and of public understandings of science, we also bring conceptions of audiences and publics together as resources for analysing science and society dynamics. In adopting such a method we are following other media and science communication scholars such as Richard Holliman who analysed cloning in relation to a circuit of culture that included production, content and reception (Holliman 2004). However, our work looks at developments in the mediations of human cloning in the decade post-Dolly, whilst Holliman's work focuses on the two years of the cloning of Dolly and the emergence of her story – 1996–97.

In much public understanding of science and public engagement literature, as the labelling of these fields suggests, non-scientists are imagined as 'the public'. As such, the public is often invoked as a generalised grouping that is assumed to constitute the audience for science communication, the investors in biotechnology, the lobbyists for or against science policy, the users of the products of science as patient/consumer groups, and as potential future scientists. As such, this general public is often assumed to be capable of particular forms of agency in relation to science. This may be envisaged as the capacity to utilise or demand it or to benefit from it. Often this is construed as a negative agency in relation to science, identified with providing insufficient funds for it, lobbying against it, failing to realise its importance and/or generally misunderstanding it. Scientists, government officials and some academics who emphasise this version regard the public as lacking understanding of science and as consequently failing to support it. However, they sometimes contend that it may be possible to transform these relations through increased and improved engagement with science.

In this use of what is often referred to as 'the deficit model' of public understanding of science, science is represented as precarious and vulnerable to damage by uninformed public opinion. Education and/or public relations have generally been regarded as the prime vehicles to tackle public ignorance or misunderstanding and thereby to render science less vulnerable. It is public support that figures as the foundation for social activity and social value in the democratic imagination. The articulation of the fear that science is lacking this support has intensified since the 1990s in the UK and the USA.

In the 1990s in the UK there was considerable concern that public trust in science had waned and consequent discussion about this constituting a 'crisis'. This understanding of the problem stimulated government investigation and evaluation (House of Lords 2000). There was a relatively large investment in public engagement and communication activities in the wake of the House of Lords Report *Science and Society* (2000) and of the general sense of crisis which had precipitated it.[12] Infrastructural support contributed to the development of what could be seen as a 'public engagement industry' in the UK during the 1990s.

The perceived crisis over public trust in science in the UK, which precipitated the consultations around and investment in public engagement activity, is usually attributed to BSE and the controversies surrounding GM crops. The subsequent reporting of the foot and mouth disaster and the measles, mumps and rubella vaccine (MMR) controversy (allegedly linking autism and the child immunisation programme) followed these developments, but these occurred after the 'crisis' had been identified. In policy discourse the reactions to BSE are often occluded by a focus on GM and this tends to foreground problems in media coverage or PR failures, rather than issues of economic or political mismanagement or benefits. Thus, the so-called crisis in trust is often regarded as a media and communications problem, rather than a problem of science policy or social organisation (Irwin and Wynne 1996).

The focus on the media as both the cause of the decline in trust in science and also potentially part of the solution to this crisis was most apparent in the establishment of the UK Science Media Centre in 2002 (see Chapter 1). In the UK there has been a substantial increase in financial and institutional support for public engagement activities since the 1990s. Science participation and consultation events such as citizen juries have flourished in the UK and mainland Europe during this period. There has also been an extension of science programming on radio and television networks. The emergence of particularly 'media-friendly' scientists such as Robert Winston (*The Human Body* (1998), *Child of Time* (2000), *Cloning the First Human* (2001)) and Armand Leroi (*Human Mutants* (2004), *What Makes Us Human?* (2005)) has provided another dimension to scientific engagement. Against this background, the UK docudrama *If. . . Cloning Could Cure Us* (2004) was thus a key marker. It was, in many respects, the culmination of this movement in that it enacted not just a technoscientific narrative, but it became a form of public engagement itself through the dramatisation of a citizen jury and the use of the telephone poll (as an exercise in consultation on science regulation). The production of *If* illuminates the underlying assumption that these practices of public engagement can be both conducted through and represented as programming content and entertainment.[13]

Deliberation and participation activities – modelled on the legal jury system – have become important initiatives to realise and extend public engagement with science. Activities described as deliberative and participatory have been conducted, formally and informally, all over Europe in the last decade, and were already in place through the *GM nation?* debates. The logic and rationale of these stagings is that they aspire to identify science with transparency and the democratic process. These engagement activities contrast with other methods that are used to gauge public understanding of science. They contrast particularly with the large-scale surveys that are used predominantly in the USA, although these have also been used in Europe (for example, Eurobarometer). Such surveys construct and mobilise rather different versions of the public.

The problems with participatory activities have been well rehearsed (House of Lords 2000) and are themselves now subject to European-wide review and evaluation. One of the main criticisms that has emerged concerns the limited social range of such activities. For example, there is widespread perception that these initiatives involve 'the usual suspects' – that is, that the participants are generally self-designated stakeholders, predominantly from the corporate world, lobby groups and some patient groups. Another criticism is that many of the so-called public consultations have not been widely or adequately advertised: as calls for participation or contributions are often 'buried on websites' and only a few people respond. Moreover, there is widespread scepticism about the results of such exercises, since some critics feel little confidence that public responses or input are taken seriously. Conversely if there is visible participation and the results are not those desired then consultations can be framed as 'hijacked'.[14] Of course, there is a range of over-arching social and political questions pertaining to how diversity, difference and multiplicity of views and positions are approached and handled in these contexts. These include: Is consensus building always possible or desirable? If there is an opportunity to make a wide spectrum of conflicting voices intelligible, how can they be assimilated into policy?

The announcement of the nearly completed human genome project (2000) and the '50 years of DNA' celebrations (2003), which were both UK–USA collaborations, contributed to the escalating public relations and publicity around human genomics at the beginning of the twenty-first century. At this juncture, the double helix had become pervasively iconic (Nelkin and Lindee 1995; van Dijck 1998) and there was much heralding of the 'genomic era'. The investment in public engagement activities complemented and accompanied the prior and ongoing investments in biomedical research, specifically around genomics, stem cells and therapeutic cloning. Moreover, attention was focused on the prospects for the UK becoming a competitive global player in these arenas.

During the first decade of the twenty-first century, the UK biomedical sciences have continued to attract funding and endorsement and a plethora of new institutions and initiatives have been launched during this period. With the Wellcome Trust, the Arts Councils, the Biotechnology and Biological Sciences Research Council (BBSRC), the Medical Research Council (MRC) and the Economic and Social Research Council (ESRC) all providing funding, in conjunction with other opportunities emerging from Europe, Canada and the USA, the resources for public engagement and science communication initiatives around human bioscience and medicine were plentiful. The Wellcome Trust's funding (the trust currently has an estimated UK annual spend of £450 million) and the ESRC's spending post-2000 on genomics, generated jobs and projects across this area in the UK and beyond.[15] These funding streams also provided opportunities for new institutional collaborations between art and science through the Sci-Art programme[16] established by the Wellcome Trust for the biomedical sciences. These institutional collaborations also include the installation of artists in residencies across the UK ESRC Genomics Network, in The BIOS Centre at the London School of

Economics, and key institutions such as the Sanger Centre and Guy's Hospital. These activities are also related to a more global proliferation in collaborations between the biosciences and the arts including centres of activity such as Symbiotica in Australia, The Arts and Genomics Centre in the Netherlands and The Art Institute of Chicago in the USA.

In April 2006 the Wellcome Trust organised a conference entitled 'Engaging Science: Is public engagement engaging the public?' This event, hosted in Manchester (UK), provided an opportunity to showcase the activities in public engagement sponsored by the Trust and other funding bodies. At this conference it was evident that there had been a shift from earlier approaches to science and the public, with the consolidation of a focus on the *processes* of public engagement itself. The presentations at this conference highlighted three main channels for public engagement:

- formal education – activities with schools and school groups, usually in schools and built into the curriculum under biology or general studies programmes;
- informal education through science centres, museums, theatres and galleries;
- mass media communications – TV, radio and the press.

Such initiatives are sustained by an imagined vision of democracy that posits that people become informed by activities in various formal and informal, public and private spheres, and they are thus able to deliberate and vote. Seen in relation to this imagined process, public engagement activities appear to be attempts to enrich the normative public sphere. From a Marxist perspective, they could be viewed as active efforts to enrol people in ideological state apparatuses. At its 'Engaging Science' conference, the Wellcome Trust described its ambition with regard to public engagement projects as follows: 'our aim is to engage the public to foster an informed climate within which biomedical research can flourish'. Thus, the flourishing of (biomedical) science, and the protection of the investment made in this field has been the primary aim of such public engagement initiatives. At this conference the emphasis was on the role of the media in contributing to this vision of democracy and the efforts to increase engagement with science. This emphasis was sustained by the presence of programmers and scriptwriters from the main terrestrial television channels in the UK.

The engaged public as it is imagined in relation to these activities can also be imagined as a set of audiences. A turn to audiences is evident in the current literature produced by the Wellcome Trust, such as the conference report for the 'Engaging Science Event' (2006). At this event, discussion of audiences became the dominant mode of talking about the groups who were 'being engaged', although the rhetoric of publics remained the meta-discursive framework.

Despite the reflexivity of much public engagement activity, the media are still often construed in this context as a kind of mass public address or news system that might reach an imagined 'general public'. However, the mass media are produced and consumed in relation to specific audience groups. For example, each broadcasting channel is divided up into specific genres such as drama, documentary and news. Media producers conceptualise their audiences in relation to specific demographics. Information about audiences feeds back into media production as considerations of viewing figures, market research and audience responses, and feedback from pre-screenings are incorporated into production processes.

Science and policy commentators refer to a media which ostensibly is as diverse and heterogeneous as 'the public'. However, the media invoked in these contexts is usually the news media and, more specifically, the press (for example, House of Lords *Science and Society* report 2000). This representation of the media as a homogenous process of news writing lends itself to the assumption, still prevalent in some versions of science communication, that the media operate through the formal model of 'sender – message – receiver' (Shannon and Weaver 1963). This version identifies the sender as the controller of the meaning of the message. This model of the media and of audiences is a transmission model of communication, evoking the image of the megaphone addressing the crowd, or the hypodermic syringe injecting the body politic.

In fact, seen through the multilayered processes of mediation and the structures of the media industries, the public is imagined as a diverse set of intersecting audience groups who select, use and remake media texts. Audiences are imagined by producers in relation to demographic information and statistics pertaining to viewing and consumption. Media producers require audience attention to ensure the purchase and consumption of (their) media texts. Science is one source of programming material or content strand, among many. Science promoters are thus in the business of trying to enrol media producers and in trying to get them interested in science (Nelkin 1987), whilst media producers borrow from science, if they feel it will help them gain audience attention (Franklin et al. 2000; Wood 2002; Hills 2004).

At the Wellcome Trust conference discussed previously (2006) one contributor on a panel about science and art collaborations commented that borrowing from or using science was a way of gaining new audiences for dance. Hence, this arts practitioner saw science as a way of making a marginal art more accessible, more public. However, the prevailing rhetoric is that in science–art collaborations, the art makes the science more accessible to audiences. The artists in science–art collaborations whom we interviewed made this assumption explicit when they articulated fears that they were being positioned as translators of science. Whilst some artists involved in such collaborations felt that this was an appropriate role for artists (and for them), several were trying to avoid becoming science communicators. These artists indicated that they were trying either to provide critical perspectives on technoscience or to encourage or provoke reflection about developments in science amongst the public. Thus, it is evident that science communication and public engagement actors may have diverse standpoints and motives. Not everyone involved in such engagement with science activities necessarily regards it as their priority to contribute to 'a climate in which biomedical (or other scientific) research can flourish'. Such activities are also about the development of media and art forms in themselves. Moreover, they may also be about providing space for critical reflection about science and techno-scientific developments, rather than operating in the mode of transmission of knowledge imagined by science communicators.

Publics as arbiters of objectivity

The question of the constitution of the public, and what is at stake in that constitution, is central to our study. The public of 'the public understanding of science' or 'public engagement with science' is often conjured as a homogenous object. Moreover, this public also emerges from a process of othering – through the production of an imagined assemblage – 'out there' – of those who are not (supposedly) already involved in science. Thus, publics are defined by absence and surplus. They are not scientists, policy makers, politicians, media producers or academics, or even informed citizens: they are everyone else. They do not belong to institutions, they are the 'new wild' of discourses of democracy and globalisation and, as such, they become at once another object (Franklin 2006) and an ephemeral multitude (Hardt and Negri 2001) in which agency is fluidly and non-specifically located. The public becomes an object that exists 'out there' beyond the institutional infrastructures of science, government and the media.

The public may become an abstracted reference point used to signal and claim legitimacy in a variety of discourses. In these representations, the public is an inexhaustible resource, the location of meaning and the objectively real. This public is reached through mobilisations of forms of representation that already exist in the performance of representative and deliberative democracy. It is reached and is accessible either through small-scale citizen juries, focus groups, stakeholder groups (all of which are have been employed in Europe), or in large-scale surveys and audience research, which are constructed to represent the opinion, desire and will of 'the people'. Thus, the public is either an inexhaustible and infinite multitude or embodied in selected representations of itself; it is either homogenous or heterogeneous and identified with the undifferentiated mass or the individuated unit. In either the close-up or long-distance focus, the matter of the public disappears into the multitude or the individual.

We do not offer an answer to the question of what the public is. Instead we raise questions about the public, science and the media. We try to make visible the complexity of the construction of these objects and subjects which are indefinable and which are seen to be crucial in the imaginings of democracy. In the recent discourses of cloning the public has been produced as a complex subject/object. It has been represented as providing a new kind of objectivity (Barthe and Lindhardt 2006) or way of providing scientific value which is implicated in a vision of a new kind of 'unbiased' subject. The public which is the subject of cloning discourses is the new arbiter of objectivity – the public in whose name cloning comes to be a scientific practice for 'future generations'. This public is composed of agential subjects, rational deliberators and those who rationally espouse a desire for cures and enhancement.

The 'other' subject is the non-subject, the 'bad' public which harbours irrational, science-fictional, misinformed understandings. It encompasses those who have irrational desires and who are beyond the reach of rational discourse and who need to be enrolled, engaged and educated. However, these 'bad' subjects do not provide new kinds of objectivity. In fact, they are rendered 'outside' the participation framework for engagement and communication, since the new forms of scientific objectivity do not allow refusal. In the paradigms of public engagement and consultation, the public is rhetorically mobilised to define the conditions of cloning developments and thereby to provide an objective basis for the conditions of research. It is precisely these exercises in engagement which also close the space for refusal.

The recent public consultation on sourcing eggs for research into therapeutic cloning in the UK exemplifies such closure. As already discussed in Chapter 5, the HFEA conducted a public consultation in 2006 on the issue of sourcing eggs for somatic cell nuclear transfer which invited responses. Despite this invitation, this consultation document demarcated the limits of engagement. The question posed to the 'public and interested parties' (HFEA 2006: 2), was: 'What conditions would make sourcing eggs acceptable?' This consultation was nested inside a set of other frames which made it impossible to participate without accepting this framing. In this exercise, the subjects of the cloning discourse were those who could speak within the established terms. The conditions for engagement were set in the multiple choices which included: first, the spectre of a fully commercial egg market; second, the compromise of a partially rewarded egg 'donation' system, in which subsidised IVF treatment would be provided for those giving eggs; and the third option of 'altruistic donation' in which rewards of any kind would be forbidden. This limited framework for participation constituted a public that could then be used to demonstrate an objective resolution – that is, public opinion is X, therefore X will be implemented.

However, the conditions for experimentation had already been established prior to the consultation. As we explored in Chapter 5, the HFEA had already licensed egg sourcing through offers of reduced-cost IVF treatment at the Newcastle Centre for Life. The HFEA did indicate that it would retract this licence should the consultation yield opposition. Nevertheless, the prior granting of a licence and the normalisation of this arrangement within techno-scientific practice overshadows the consultation exercise and sets up a normative expectation that the practice has already been deemed acceptable.

The public mobilised in this consultation is the public which operates as subject to the discourses of cloning, materialising in bodies that will provide eggs and in bodies which will fill in the consultation forms. This same public subject is called upon to be or produce the bodies 'in need' of stem-cell cures. Cloning has been developed as reproductive through mammalian cloning, but the focus for recent development has been therapeutic human cloning. As we have shown, this technoscientific practice is closely linked to reproduction, derivative of IVF, reliant on eggs and performed by fertility clinicians such as Alison Murdoch. Adele Clarke has observed that:

> in the reproductive arena, there is no neutral position, and most people are aware of this. Alternatively, in science studies terms, we might state that in the case of reproduction there is no boundary between science and society (Gieryn 1999). In fact, the reproductive sciences have been built upon the transparencies of both social life and their own technoscientific products. Political challenges and moral crusades abound and will continue.
>
> (Clarke 1998: 235)

As Clarke's comments indicate, reproductive science and biomedical research are always clearly implicated in the social world. Their practices cannot be designated as 'basic research', which is discursively constructed in science as non-commercial or as not focused on application, as they are clearly already clinical and because they incorporate the body within the practices of science. Moreover, contemporary scientists are aware that while 'basic research' may remain an ideal of 'Science', it is not sustainable in the context of contemporary funding regimes because it can neither be sold nor provide social intelligibility. The fact that the reproductive sciences cannot be effectively positioned as having a 'basic research' component means that the mobilisation of the public in relation to them is particularly powerful and important. This mobilisation represents a moral authority which becomes, in effect, a supplement to objectivity. In these sciences (rational) public opinion, through its offering up of bodies, operates as the justification for proposed (and, sometimes, already established) practices. Such bodies operate through an 'ontological choreography' (Thompson 2005), sacrificing one kind of subjectivity for the opportunity of transformation. As we argued in Chapter 5, women's bodies operate as the boundary figures through which cloning as a respectable and desirable scientific process can be articulated. Likewise, claims to know the public operate as legitimating practices which provide new arbiters of objectivity and the basis for public decision making about science policy and research.

Conclusion: audiences for the democratic imaginary

The democratic imaginary is constituted though multiple visions of community, communication and participation. All the sites in which cloning is consumed are constitutive of the public understanding of cloning: fiction and non-fiction, news and film, art and science. We have argued throughout this book that there are dominant discursive frames and strategies which shape understandings of cloning. We have also illustrated some of the ways in which these dominant frames become subject to revision, contest and challenge through alternative imaginings and practices. These dominant discourses are often iterated and reiterated through institutional mechanisms and through the interventions of relatively powerful actors – including some scientists, politicians and journalists. Alternative imaginings are often positioned as marginal and construed as 'anti-science'. It appears that there is no space to refuse scientific innovation once it is framed as such. The very articulation of a process as scientific seems to guarantee an ontological purchase not available in other spheres.

What is at stake in the construction of publics and audiences in cloning discourses? At stake in these constructions are the prospects for democratisation of science and expectations about the role of science in the democratic imaginary (Laclau and Mouffe 1998). These imagined prospects involve making science as a set of institutions more democratic, and also offer the prospect of science playing a greater role in the constitution of democracy. In the dominant liberal democratic imaginary, sufficient information about science allows the people, in whose name power operates, to deliberate and arrive at scientifically informed norms and through this the democracy may be sustained. However, public engagement is also about the interpellation of the political subject into a specific democratic imaginary. In the biopolitical context of cloning, entering into a discourse supporting therapeutic cloning is the qualification for being an intelligible liberal political subject. The call for enhanced public engagement with science in the early twenty-first century, identified particularly with the House of Lords Report (2000) in the UK, was also a call for a more effective and scaled-up interpellation exercise. On the one hand, public engagement offers the prospect of wider participation in deliberation about science. However, on the other hand, it is simultaneously a colonising move which expands the category of rational political subjects, in the face of fears of the irrational masses, and secures cloning as a rational political project in the UK and elsewhere.

8 Conclusion

Mention of either cloning or the media tends to elicit generic responses and fairly sweeping declarations. This conclusion addresses some of these standardised evaluations head on by drawing out some of the key arguments and reflecting on the features of the analysis offered in this book.

'It's all science fiction!'

In the early 1990s many commentators assessing the prospects for human cloning could offer this appraisal with confidence and impunity. However, in the wake of the announcement of the arrival of Dolly, the cloned sheep (1997), human beings faced 'the possibility of human cloning' (Nussbaum and Sunstein 1999: 11). By 2005, prestigious scientific journals – most notably *Nature* and *Science* – were announcing dramatic breakthroughs in this field.

This book has investigated this transformation, tracing the changes in the meanings of, and expectations for, human cloning that have been realised at the end of the twentieth and beginning of the twenty-first centuries. We have shown that, during this period, human cloning was transposed from the realm of science fiction to that of technoscientific practice. This has been a remarkable transformation and the project of this book has been to trace the cultural dimensions of this shift: to show what has been involved in the making of human cloning into a technoscientific practice.

In effect, we present a story of human cloning as it emerged in the late twentieth and early twenty-first centuries. Of course it would be impossible to tell this story without an awareness of what preceded it. Hence, we set out to investigate the history of human cloning prior to 1996. In so doing we became aware of the many stories and genealogies of human cloning that had been assembled, particularly the flurry of accounts that emerged in the wake of the cloning of Dolly the sheep. We became intrigued by the specificity and partiality of these ways of telling the (hi)story of cloning and by what these genealogies included and what they left out. Chapter 2 provides a critical reading of some of the most prominent accounts that have appeared since 1996. While these criticisms are important, our main intention is to highlight the complexity of human cloning as a cultural phenomena and to trace its rich, if tangled, history.

This book shows that human cloning is a complex cultural figuration that has been made and continues to be made in many different sites. This is exemplified in our excavation of its multi-layered relationship to the so-called 'new reproductive technologies' of the late twentieth century and to IVF in particular. While Ian Wilmut and Roger Highfield, as well as Gina Kolata, have acknowledged some features of this strand of cloning genealogy, their perspective is narrowly technical. As we indicate, they draw attention to the development of the skills of micromanipulation in late twentieth-century human infertility treatment and, specifically in IVF, as providing some of the expertise and skills that led to Dolly. We identify other threads in this legacy, particularly in the UK. We have shown that, in this national context, the legal framework forged in the new reproductive technology and embryo debates of the late 1980s underpinned developments in technoscientific cloning a decade later, facilitating the embryo experimentation that cloning entails. Moreover, as Karen Throsby (2004) has argued, the normalisation of IVF has opened the way for other forms of experimentation in genomic and reproductive technoscience and made such practices seem both feasible and desirable. Michael Mulkay's (1997) account of the transformation of expectations around embryo experimentation from a 'rhetoric of fear' to a 'rhetoric of hope' realised in the parliamentary debates concerning NRTs during the late 1980s provided us with a template for tracing a similar transformation in expectations regarding human cloning at the turn of the twentieth century. Finally, we have focused on the material and institutional dimensions of cloning technoscience to draw out its links with IVF. The normalisation of IVF has quite literally made women's bodies available to resource therapeutic cloning. The location of regenerative medicine centres adjacent to IVF clinics materially and symbolically signals this. Recent calls in the UK for women to donate their eggs for therapeutic cloning experimentation in exchange for cheaper IVF treatment or 'altruistically', receiving only compensation for lost earnings, both mark and solidify this genealogical trail (cf. HFEA consultation 2006).

Our critical scanning of missing trails in established accounts and our proposed additions to these provide the background for our main investigation which concerns human cloning in the late twentieth and early twenty-first centuries. The changes which we have traced are subtle but significant and, to a considerable extent, they are still being enacted. Nevertheless, our task has been to identify the key cultural elements in this transformation and to outline some of the ways it has been achieved and continues to be made. We have highlighted three major changes in the figuring and meaning of human cloning which have been crucial to its fabrication since 1996. Thus, we have tracked: the shift in expectations around cloning as these were increasingly focused on the promise of imminent cures; the change of focus from the figure of a fully cloned human subject to the micromanipulation processes of cloning; and the altered iconography of human cloning as the image of the enucleation of the embryo replaced the image of the multiple clone.

With explicit reference to the work of other scholars, we trace the emergence of the distinction between therapeutic and reproductive cloning at the end of the twentieth century. As we have insisted, this is by no means a natural distinction, given that reproductive and therapeutic cloning involve the same original technological processes. Nevertheless, this prying apart of these forms of cloning has been, and continues to be, a significant rhetorical move. Our emphasis has been on the discursive work this requires and on the considerable investment in it, particularly in the UK context. The analysis of the making of this distinction is a key strand in our research and we have shown how it severs cloning from its material and embodied moorings in, and associations with, human reproduction and how it obscures the corporeal context of women's bodies. Furthermore, we have stressed that the othering of reproductive cloning has symbolically detached therapeutic cloning from the horrors and fears which have predominated in science fiction and associated it instead with the 'rhetoric of hope' and the promise of cures.

The distinction between therapeutic and reproductive cloning has been maintained and reinforced, to a considerable extent, by the enormous investment (in all senses of this term) in the curative potential of cloning science. The re-casting of human cloning as therapeutic has been a striking feature of its late twentieth and early twenty-first century incarnations. Indeed, as we have noted, the term 'therapeutic cloning' declares what this technoscientific process will achieve and assumes that achievement. The ever-extending list of diseases and conditions designated for cloning therapy is perhaps the most obvious marker of this casting. Our study has identified many other vehicles and sites of this cultural rendering of cloning as therapeutic. These have included the inscription of celebrities seeking cures or campaigning in support of stem-cell or cloning research (see Chapter 3) as well as the interpellation of audience members as seekers of cures in the addresses of celebrated scientists. In this field, further celebrity engagement (particularly in the USA) as harbingers of and spokespersons for the promise of cloning cures may be expected.

We have emphasised throughout this book that extensive discursive work has been required precisely because the distinction between therapeutic and reproductive cloning is a fragile one. We have noted moments and locations (particularly in film) in which it has been blurred or breached. We have also highlighted its temporality. So, for example, while there has been considerable disavowal of reproductive cloning since 1997, few scientists in the field would deny that it may be undertaken at some point in the foreseeable future. Moreover, we have pointed to particular recent films and some specific policy contexts (in the UK) which suggest that reproductive cloning, particularly through its connections with therapy (as a cure for infertility), may be gaining increasing legitimacy.

'It's all media hype!'

In the course of our research for this book, we were amazed to encounter the familiar refrain 'media hype' repeatedly in discussions of bioscience. Many rigorous and insightful social science researchers had no compulsion about brandishing it, in either formal or informal presentations and discussions. This sometimes occurred as they interrupted the flow of their sophisticated analysis or thoughtful reflection on aspects of bioscience to offer this sweeping assessment. But it is not only academics undertaking social studies of bioscience who castigate or dismiss the media in this way. Scientists also criticise the media for framing the cloning debate inappropriately. Moreover, we have shown that even some recent initiatives to engage the public with science in the UK have replicated this pattern of blaming the media as the public is construed as misinformed (see Chapter 7).

If there is one message that we hope to convey with this book, it is that it is impossible to understand twenty-first century bioscience without taking the media seriously. Hence, this book confronts and challenges the ubiquitous cliché cited above. We have stepped back from dismissive assessments of the media and from indiscriminate 'media bashing' to explore *why* and *how* mediation matters in twenty-first century genomics and in cloning science in particular. This book shows in detail some of the ways in which the media have been actively involved in making the technoscience of cloning, and it foregrounds several dimensions of the mediation of cloning technoscience.

We have underscored the constitutive role of such mediations. For example, we note the significance of declaring national investments on an international stage and of affirming national stakes in this field. Through some detailed analysis, we have shown how therapeutic imagery has generated new meanings and expectations around cloning. Working with an unusually wide variety of texts and across a range of media, our research extended beyond conventional media sites as we encountered other cultural forms and processes that have been crucial to cloning technoscience. These range from a national postage stamp to key policy documents and scientists' own accounts of their work. We have also analysed public engagement events, consultation documents and materials emerging from our own audience research initiatives. We have explored the specific features of these diverse cultural materials and processes and the links and gulfs between them: their resonances and dissonances as they contributed to the making of human cloning.

Moreover, we have analysed accounts of scientific developments written by leading scientists alongside film or newspaper reports. Such methodological stretchings of the parameters of media analysis have been quite deliberate: they have been undertaken to challenge explicitly the conventional division between science and mediation. In this respect, our research practice is consonant with our theoretical and political insistence that there are no such things as unmediated scientific facts or neutral accounts.

While our extensions of research parameters have been principled, we recognise that these raise methodological and conceptual challenges which merit some explicit address. We were striving for what might be characterised as methodological symmetry, in that we set out to treat the diverse cultural products we analysed similarly and critically (Bloor 1976; Latour 1986). However, we have also paid attention to specificity of form. In working with a scientist's account of how cloning developed, for example, we take this text not just as source material, but also as, in specific contexts, a media-mobilising discursive strategy. Similarly, in examining an HFEA consultation document regarding egg donation, we were as attentive to its language, images and its framing of the debate, as we were when analysing newspaper articles or films.

Our symmetrical approach to diverse cultural texts was complemented by our concern to register the *context* in which such different materials are produced and the audiences for which they are intended and to note their *differences*. We explore, for example, how the status of different genres inflects (and is inflected through) the debate about human cloning and we note the different processes of 'gate-keeping' through which diverse representations have to pass (whether that is peer review for a scientific paper, or the evaluation of commercial potential for a research enterprise, a patent, or for a mass-circulation film). We were also interested in the values (for example, 'news values') and the material resources which shape and underpin diverse cultural productions (from state funding for major research initiatives to corporate funding for a Hollywood film).

Attention to such differences has worked alongside the investigation of how tropes travel across different media forms, genres and outlets. In tracing the flow of representation across different genres, we acknowledge differences but also recognise the importance of not treating spheres of representation as hermetically sealed-off from each other. A film such as *Godsend* (discussed in detail in Chapters 5 and 6) is a corporate product which goes out through established distribution outlets and may be presented in commercial cinemas or viewed as a DVD or video in more private forms of consumption, or be used as a teaching aid in public engagement with science events. It is cross-referenced in multiple texts (including other films and news reports). We have also observed how it has been extended into a virtual marketing strategy through the web-based promotion of The Godsend Institute, which claimed to offer reproductive cloning (discussed in Chapter 6). The movements across 'fact'–'fiction' divides have been brought under scrutiny, as have some important moments of interplay between cultural representation, political policy and scientific practice.

This study identifies ways in which the mass media are increasingly being incorporated into the daily operations of early twenty-first century biosciences. For example, we looked at how mass media coverage of particular fields can facilitate communication between scientists in these fields as well as inter-specialist communication to other scientific communities. One consequence of this is that the mass media may enable problems (including fraudulent work) to be identified more quickly. The role of one television programme and the internet in opening Hwang's work to both scientific and broader social scrutiny demonstrated this. In addition, our study has traced some examples of the complex boundary work undertaken in and through the mass media to designate legitimate members of the scientific community (see Chapter 5). As we have indicated, the mass media have become crucial in securing funding, ensuring legitimacy and garnering support for enabling legislation and regulation in the biosciences (and other fields). Our explorations of the instantiation of a 'rhetoric of hope' (Mulkay 1997; Chapter 2 this book) and 'discursive overbidding' (Nerlich and Clarke 2003; Chapter 3 this book) have fleshed out some of the detailed ways this has worked in recent cloning technoscience.

This study has also reflected on the growing awareness of the importance of the media amongst individual scientists and the national scientific community in the UK. The establishment of the Science Media Centre in London (2002) is one obvious institutional signal of this. The emergence of celebrity scientists including Robert Winston and Ian Wilmut is another marker of scientists' media engagement. Of course, this has also been encouraged through the programmes and initiatives around science communication and public engagement sponsored in the UK by government, the Royal Society, the British Association for the Advancement of Science and particularly the Wellcome Trust as discussed in Chapter 7.

However, in this book we have also drawn attention to some of the more subtle and striking manifestations of this intensified involvement with the media. Debates about scientific witnessing have, to some extent, become public debates and this includes controversies about how the media should participate in this process. So, for example, we have observed the peer-review system coming under the scrutiny of a wider public through Richard Horton's (2005) letter to the *Guardian* in the wake of the Hwang scandal. Robert Winston's (2005) critical assessments, raised in a meeting of the British Association for the Advancement of Science in 2005, about scientists taking responsibility for managing public expectations in relation to stem-cell research, is another interesting example of developing scientific reflectivity about the media.

Finally, we have set out to unpack conceptions of the public, using some of the tools of audience research. We have suggested that such resources may be helpful in identifying the limits of some recent initiatives in public engagement with science and in developing more fully democratic frameworks for public understanding of/public engagement with science. Our analysis here shows ways in which participation is restricted by rationalistic and consumerist framings of publics.

'As an asexual form of reproduction, cloning undermines the established heterosexual order'

Cloning has long been imagined as an alternative to heterosexual reproduction and for this reason it has either been positively anticipated as potentially liberating or considered threatening, associated as it is with the spectre of excessive sameness and homosexuality. Jackie Stacey's research on the late twentieth-century and early twenty-first century cinematic representations of genetics has explored this imaginative terrain. Our research, like Stacey's, indicates that these apocalyptic visions of revolution or disturbance in the gender/sexual order have continued to animate cinematic producers, narratives and viewers (Chapter 5 this book; Stacey 2003, 2005, forthcoming).

Nevertheless, in general, our deciphering of the hidden and sometimes occluded gender and sexual relations of human cloning in the early twenty-first century reveals a more familiar pattern which is disturbing in a rather different way. The representations of this technoscience, particularly in press and television news coverage, have generally been firmly anchored through conventional binary gender codings and heteronomative framings. While professional cloning science has not been 'a world without women' (Noble 1992), our brief analysis of media coverage of Alison Murdoch from the Newcastle Centre for Life and of Brigitte Boisselier of Clonaid demonstrates that women do not fit easily into the picture of this technoscience (see Chapter 4). We have also been troubled by how women's bodies appear and disappear in recent cloning discourses and by the selective strategic inscription of women into the world of cloning and stem-cell technoscience, most notably as carers and egg-donors (Chapter 5). We have indicated that some alternative films have afforded more imaginative and utopian explorations of this dimension of cloning technoscience. Moreover, there has been some political protest both in South Korea and in the UK about protocols and procedures to secure eggs from women for stem-cell experimentation. However, there seems little prospect that technoscientific cloning will disturb established gender regimes. Moreover, there are grounds for concern about how this technoscience may be implicated in other vectors of social inequality – particularly given global markets in gametes and the stratifications in access to (in)fertility treatment.

Our charting of the changing visions of the future for cloning (see especially Chapter 3) has shown that cloning has accrued a diverse set of expectations in the twenty-first century. Therapeutic cloning is increasingly regarded as a technoscientific vehicle for improving individual health and extending longevity. Strikingly, its therapeutic capacities are now sometimes seen as extending into reproduction in a somewhat different way as it is envisaged as potentially dealing with fertility – as a cure for infertility. In this sense, recent cloning discourse constitutes this technoscientific practice as multiply regenerative, rather than asexually reproductive.

'The Human Genome Project, genomics, cloning: it's all global science'

This project examined the making of technoscience, the making (and policing) of an international scientific community and the making of publics who can appropriately engage with this technoscience. We have sketched some of the ways in which this making has been realised in and through the global flows of early twenty-first century information and communication technologies. The Hwang affair has been a prime example as this South Korean scandal about scientific practice and governance became a scientific crisis on a global scale. Likewise, mass-circulation Hollywood films have circulated a distinct repertoire of images of cloning internationally. The features of this imagery have preoccupied us in this study. We have also registered the casting of the global subject (Haraway 1991) who is the anticipated beneficiary of the curing promise of cloning technoscience.

We have also contended that the 'imagined communities' (Anderson 1983) of particular nation states have been recast and are being consolidated in and through recent developments in cloning and genomic science. This has been particularly apparent in the UK, where Tony Blair has endorsed this field to make it a centrepiece in his vision of the nation. This was obvious from his joint announcement with Bill Clinton of the 'completion' of the Human Genome Project (HGP) in 2000, through some specific speeches (see Chapter 7) and his recent visit (2006) to California to develop collaborations with stem-cell research initiatives there. But Blair is a celebrated spokesperson for a more general effort to stake the UK's claims in the field of biotechnology. Much has been made of Britain's identification with key achievements in reproductive, cloning and genomic technoscience. Indeed, such achievements are often cited as signs of the national promise of this field, including: Britain's leading role in IVF experimentation and institutionalisation, Dolly the sheep and the HGP. Beyond this and less obviously, we have shown a wide array of discursive inscriptions and interpellations through which the UK public is being enrolled in this national vision. Moreover, Britain's bid to become a world leader in biotechnoscience has been consistently framed in terms of its capacity to sustain such technoscience within a framework of rational regulation and governance. In this context, our research on the long-term significance of Britain's early debates on embryo experimentation and new reproductive technologies, on the HFEA and its consultation and on recent initiatives about public engagement with science signal this casting. We are rather suspicious of this specific mobilisation of national vision, since it seems to draw on an imperial legacy based on claims of administrative competence and superiority. We are also concerned about the restrictions it seems to be placing on the prospects for democratising science – as it is being enacted through contrived consultations and limited public engagement stagings.

The picture in the USA has its own complexities and our research on this national context has been more limited. However, we have noted that the Blair–Bush alliance has not extended into the field of biotechnology, cloning and stem cells.

President Bush's embargo on this field does seem to have instigated something of a backlash (particularly in California). In the USA, it has been predominantly Hollywood celebrities, rather than celebrity scientists, who have been the public advocates for therapeutic cloning and stem-cell science.

It is hard to predict how the combination of a highly commercialised bioscience sector, the escalating demands of wealthy health-care consumers, and Christian conservatism will play out here. However, since the USA continues to dominate the global market for popular English language film production, whatever the government stance on human embryonic experimentation, it will remain an important location for the conjuring of genomic imaginaries.

For a brief period in the early twentieth century, South Korea was positioned as leading the world in human cloning technoscience. We have analysed some of the modes and craftings involved in this position in the UK news media context. The story, or more precisely, the *stories* of the rise and fall of Hwang Woo-Suk have proliferated in many domains: in the alternative and the mainstream news media of South Korea, as well as in the UK and USA and elsewhere in the world; in the communication media of various scientific communities; and in publications and forums of ethics committees, science studies researchers and so on.[1] We have analysed the crucial role the Korean media played in the subsequent discrediting of Hwang and the versions of this story as it was reconstructed in the UK media. The long-term significance of Hwang's rise and fall and its consequence for South Korean bioscience are by no means clear at this point. However, we can confidently predict that the media will continue to play an important role in the unfolding of that national story and its various distinct versions in other parts of the world.

This book has been about the playing out of global and national visions of technoscientific possibilities for human cloning in and through the media. We have been able to monitor some of the global flows and national specificities in the mediation of human cloning during the period of our study. Although this has been a crucial dimension of this research project, it is certainly an area which demands ongoing monitoring and further investigation.

After clichés: our ending

This book has brought together technoscience studies and media studies. It has followed and analysed the transformation of human cloning from the world of science fiction to the world of technoscientific practice in the late twentieth and early twenty-first centuries in the UK, with reference to related developments in South Korea and the USA. We have tried to get beyond clichéd framings of mediation and cloning to understand the dream machines, the discursive practices and the material and symbolic investments involved in making cloning technoscience. We have demonstrated that the meanings of this particular technoscience have been, and continue to be, constituted in a variety of sites. The media have not merely provided pictures of the technoscientific world of human cloning since the first pictures of Dolly the sheep in 1997, they have also been implicated in the various ways sketched in this book in the making of that world. This book has drawn attention to the power relations embedded in various processes of mediation: from the conjuring of the UK as the knowing centre of international science governance and regulation, to the address to women to identify themselves as altruistic egg-donors for cloning experimentation, to the restrictions demanding rationalistic forms of engagement which offer only the opportunity to endorse technoscience as it is. Technoscientific human cloning as it has emerged from various mediations in the twentieth and early twenty-first centuries promises much. It remains to be seen precisely what it will deliver.

Appendix 1

Cloning timeline

This table draws together a composite account of key moments in the recent and more extended history derived from accounts of developments in the field and contemporary news accounts. We have overlaid on these accounts some related historical markers in the human biosciences.

Late 19th/early 20th century	German biologist Hans Driech split the cells of sea urchin embryos and produced fully functioning single-celled embryos (two from a two-celled embryo, four from a four-celled embryo).
	Hans Spemann performed similar, successful experiments with two-celled salamander embryos.
	Experiments with parthenogenesis in sea urchins by Jacques Loeb provided important lessons about the totipotency of embryonic cells and the influence of cell cytoplasm on differentiation.
1950s	Robert Briggs and Thomas King, US embryologists, developed technique of nuclear transplantation, removing nuclei from frogs' eggs and replacing each nucleus with a tadpole embryo cell of the same species to create tadpoles that were healthy copies of their nuclear donors (1952). This raised the possibility that the differentiated cells of an adult could be used to generate a living clone.
1966	Oxford biologist Sir John Gurdon produced cloned frogs using cell nuclear transfer. He 'reported that when he inserted the DNA of a cell nucleus from a juvenile animal into enucleated eggs, he could make a frog every so often' (Wilmut and Highfield 2006: 58). The cell nucleus was obtained from a tadpole, not an adult frog.
1971	James Watson published 'Moving toward Clonal Man' in *Atlantic* magazine.
1972	Willard Gaylin, co-founder of the Hastings Centre initiated public debate on cloning in *The New York Times Magazine* in a bid to win social prominence for bioethics.

1975	Convincing evidence published in the *Journal of Embryology and Experimental Morphology* that Gurdon was using fully differentiated donor cells for the nuclei he was transplanting.
	Derek Bromhall's rabbit cloning experiments (Bromhall 1975).
25 July 1978	Birth of Louise Brown, the first child born following successful *in vitro* fertilisation (IVF) of a human egg.
1980s	Neal First and colleagues at the University of Wisconsin, working with cattle, pioneered the use of an electric current to stimulate the fusion of an embryonic cell with an enucleated egg.
1981	Peter Hoppe and Karl Illmensee claimed to have cloned mice (1981).
1984	Steen Willadsen cloned sheep embryos by nuclear transplantation (nucleus donors 8-cell embryos) (1986).
1990	Human Genome Project (HGP) officially begins in United States.
	Human Fertilisation and Embryology Act passed in UK.
1991	Human Fertilisation and Embryology Authority (HFEA) established.
March 1996	UK mass media coverage of the births of Molly and Megan at Roslin Institute. Molly and Megan were the offspring of embryos 'cloned' using cell nuclear transfer techniques. Embryonic stem cells were used to provide the nuclei that were transferred.
5 July 1996	Birth of Dolly, the first mammal resulting from the artificial creation of an embryo through the process of cell nuclear transfer using a somatic – that is a fully differentiated – cell. This 'second creation' took place almost 18 years after birth of Louise Brown.
23 February 1997	Announcement of the birth of Dolly in the *Observer*.
February 1997	President Clinton calls on US National Bioethics Advisory Panel to report within 90 days on the implications of Dolly's birth for human cloning.
March 1997	President Clinton announced a ban on US federal funding for cloning.
July 1997	Birth of Polly at Roslin Institute; a genetically modified sheep which produced Factor IX, a clotting protein for treating haemophilia in humans.
5 December 1997	Richard Seed announces his intention to do human cloning before legislation is formed.
1998	Human embryonic stem-cell lines established by researchers at Johns Hopkins University and University of Wisconsin.
	Severino Antinori, Italian-based infertility treatment provider, announces plans to use human cloning to help infertile couples.

December 1998	*Cloning Issues in Reproduction, Science and Medicine: A Report by a Joint Working Group of the UK Human Genetics Advisory Commission and the UK Human Fertilisation and Embryology Authority.*
2000	UK House of Lords Select Committee on Science and Technology Third Report: *Science and Society.*
	Draft version of human genome sequence completed.
2001	Draft version of human genome sequence published.
	Antinori and Panos Zavos together with Brigitte Boisselier give evidence at National Academy in the USA on the prospects for human reproductive cloning.
	USA House of Representatives votes to ban all human cloning.
February 2002	*House of Lords Select Committee Report on Stem Cell Research.*
2003	Finished version of human genome sequence published.
	Dolly the sheep dies.
January 2004	Zavos calls press conference to announce that he has introduced a cloned human embryo into the womb of an infertile woman.
February 2004	Hwang Woo-Suk announced the successful cloning/SCNT production of a human embryo at Seoul National University in South Korea (Hwang et al. 2004).
August 2004	UK Newcastle Centre for Life received first licence from HFEA to conduct cell nuclear replacement research (experiments).
November 2004	Proposition 71 passed in California. This proposition authorised the state of California to issue bonds to raise funds for embryonic stem-cell research.
March 2005	UK – *Human Reproductive Technologies and the Law: Fifth Report from the Science and Technology Committee.*
May 2005	Hwang and Gerald Schatten announce the creation of patient-specific embryonic stem cells through use of somatic cell nuclear transfer technology (Hwang et al. 2005).
August 2005	Hwang clones Afghan hound: Snuppy (Seoul National University Puppy).
December 2005	Hwang and Schatten retract May 2005 *Science* paper.
April 2006	USA – California Senate bill (SB 1260) (regulating egg sourcing for research).
September to December 2006	HFEA Consultation *Donating Eggs for Research: Safeguarding Donors.*

February 2007	HFEA announced outcome of *Donating Eggs* consultation: women to be allowed to donate their eggs for research projects in conjunction with their own IVF treatment or as altruistic donors. See www.hfea.gov.uk/en/1491.html for text of press statement.

The timeline is drawn from the following resources: Bowring (2004), Klotzko (2001), Kolata (1997b), Wilmut et al. (2000) and Wilmut and Highfield (2006).

Appendix 2

Television and film texts referenced in the project

Films

Title	*Date*	*Director*	*Country*	*Studio*	*Link to genomics*	*Keywords*	*Genre*
Invasion of the Body Snatchers	1956	Siegel, Don	USA	Allied Artists	Pods from outer space descend on a small US town, where they proceed to replace humans with soulless alien clones.	clone	science fiction
Village of the Damned	1960	Rilla, Wolf	UK	MGM	Women in a remote town are mysteriously impregnated with alien DNA during a power blackout, and simultaneously give birth to identical children who communicate telepathically. (Based on the novel *Midwich Cuckoos*.)	DNA	horror
Andromeda Strain, The	1972	Wise, Robert	USA	Universal Pictures	An extra-terrestrial virus, without DNA, mutates and spreads amongst humans.	DNA	science fiction
Clones, The	1973	Card, Lamar and Hunt, Paul	USA	Filmmakers International	A nuclear scientist is 'cloned' by foreign agents, struggles to prove he is himself and not the clone.	clone	action, science fiction
Stepford Wives, The	1974	Forbes, Bryan	USA	Palomar/Fadsin	All the other men in the village have replaced their wives with subservient androids (visual clones), and her turn is next.	clone	horror, science fiction
Shivers	1975	Cronenberg, David	Canada	Canadian Film Development Corporation	Genetically engineered parasites are used as organ transplants, they spread out of control amongst the population.	gene, genetic	horror, science fiction

Demon Seed	1977	Cammell, Donald	USA	MGM	A super-computer called Proteus wants to break free. It rapes the wife of its creator, impregnating her with computer-generated DNA. This leads to the birth of a man-machine hybrid.	DNA, hybrid	horror, science fiction
Island of Dr Moreau, The	1977	Taylor, Don	USA	AIP/Cinema 77	Dr Moreau creates man-beast hybrids on a remote island, who eventually rebel and kill him.	hybrid	horror, science fiction
Boys From Brazil, The	1978	Schaffner, Franklin J.	USA	ITC/Producer Circle	Dr Josef Mengele clones Hitler and puts out the boys for adoption, but his plans are uncovered by Nazi-hunter Lieberman. Is nature stronger than nurture?	clone	horror
Parts: The Clonus Horror (Clonus)	1979	Fiveson, Robert S.	USA	Fiveson Productions	Politicians scheme to clone themselves, assuring immortal life.	clone	science fiction
Blade Runner	1982	Scott, Ridley	USA	Warner	Deckard must retire a group of replicants – genetically engineered servant clones with a limited lifespan – who have illegally returned to Earth.	gene, genetic, clone	science fiction
ET: The Extra Terrestrial	1982	Spielberg, Steven	USA	Universal Pictures/ Amblin	Elliot befriends a young alien, his alter-ego, who also has DNA and who has also been abandoned by a parent.	DNA	science fiction
Star Trek II: Wrath of Khan	1982	Meyer, Nicholas	USA	Paramount Pictures	Khan and refugees from the eugenics wars of Earth, hijack the ship and steal the 'Genesis Device'.	Eugenics	science fiction

Films *(continued)*

Title	*Date*	*Director*	*Country*	*Studio*	*Link to genomics*	*Keywords*	*Genre*
Thing, The	1982	Carpenter, John	USA	Universal	An alien attacks an Arctic research station, taking over the DNA and bodies of its victims.	DNA	horror
Creator	1985	Passar, Ivan	USA	Universal Pictures	Harry wants to clone his dead wife, but first he needs an egg and a host.	clone	comedy, romance, science fiction
Fly, The	1986	Cronenberg, David		Brooksfilms	A common housefly entered the 'telepod' with Seth, and the machine spliced their DNA together.	DNA, gene	horror, science fiction
Fly II, The	1989	Walas, Chris		Brooksfilms	A few years after Seth Brundle dies. Seth's employer, Bartok (Lee Richardson), adopts Martin (Seth's son) to have the genes so he can use them to create a super army of genetic flies.	gene	horror, science fiction
Metamorphosis	1990	Eastman, George	Italy/USA	Filmirage	A brilliant geneticist is working on a serum which will stop human ageing. He administers the serum to himself, but the results are unexpected and horrendous.	gene, genetic	horror, science fiction

Guyver, The	1991	Mad George, Screaming	USA	New Line Cinema	A young man discovers a mechanical device that merges with his own body, turning him into a cyborg superhero, he begins to uncover a secret plot to genetically engineer terrifying monsters.	gene, genetic	science fiction
Johnsons, The (Johnsons, De)	1992	van den Berg, Rudolf	Netherlands	Meteor Film Productions	Brothers reproduced within the framework of mysterious genetic experiments terrorise a young innocent girl who seems to be chosen for a sinister destiny.	gene, genetic	horror, thriller
Army of Darkness	1993	Raimi, Sam	USA	Universal Pictures	Evil clone must be defeated.	clone	comedy, horror
Body Snatchers	1993	Ferrara, Abel	USA	Warner Brothers	Pods from outer space descend upon a US military base, where they proceed to replace humans with soulless alien clones.	clone	horror, science fiction
Carnosaur	1993	Simon, Adam	USA	New Horizons	Mad scientist Dr Jane Tiptree crosses chicken eggs with dinosaur DNA, and gestates the embryos in human surrogate mothers.	DNA	science fiction
Jurrassic Park	1993	Spielberg, Steven	USA	Universal Pictures/ Amblin	Dinosaurs, cloned from DNA obtained from insects trapped in amber.	DNA	science fiction, action
Lorenzo's Oil	1993	Miller, George	USA	Universal Pictures	A genetic disease is killing a child.	gene, genetic	drama

Films *(continued)*

Title	*Date*	*Director*	*Country*	*Studio*	*Link to genomics*	*Keywords*	*Genre*
Street Fighter	1994	de Souza, Stephen E.	USA/Japan	Capcom Entertainment	Creation of genetically enhanced warrior.	gene, genetic	action, science fiction
City of Lost Children (La Cité des enfants perdus)	1995	Jeunet, Jean-Pierre	France	Claude Ossard	Six clones, a disembodied brain floating in a fish tank and a mini Bride of Frankenstein also inhabit the oilrig.	gene, genetic, clone	horror, science fiction
Judge Dredd	1995	Cannon, Danny	USA	Cinergi Pictures	Judge Dredd is framed when his DNA is found on a murder weapon. While in exile, he learns that he was born as the result of a genetics programme and he has a cloned brother Rico.	DNA, clone, gene, genetic	science fiction
Pinky and the Brain	1995	Calabrese, Russell. Caldwell, Barry et al.	USA	Amblin Entertainment	Pinky and the Brain are two laboratory mice living at Acme Labs whose genes have been spliced. The Brain is a genius, whereas Pinky is quite insane.	gene	animation
Species	1995	Donaldson, Roger	USA	United Artists/ MGM	Government scientists use genetic engineering to combine human and alien DNA. The outcome is an attractive but murderous woman, Sil, who will go to extreme lengths to spread her genes within the human population.	gene, genetic, DNA	science fiction, thriller, horror

DNA	1996	Mesa, William	USA	InterLight Productions	Wessinger has genetically engineered alien DNA, from an ancient fossil, with an immunity-preserving beetle chemical, to make a monster.	DNA, gene, genetic	horror, science fiction
Island of Dr Moreau, The	1996	Frankenheimer, John	USA	New Line	Dr Moreau genetically engineers man-beast hybrids on a remote island, who eventually rebel and kill him.	gene, genetic, hybrid	horror, science fiction
Multiplicity	1996	Ramis, Harold	USA	Columbia Pictures	A building contractor has himself cloned to make more time for himself.	clone	comedy
Nutty Professor, The	1996	Shadyac, Tom	USA	Universal Pictures, Imagine Entertainment	Sherman Klump is a college professor on the verge of a breakthrough in DNA restructuring. He decides to test a formula on which he's been working and is then transformed into the lecherous swinger, Buddy Love, and romantic complications ensue.	DNA	science fiction, comedy, romance
Alien: Resurrection	1997	Jeunet, Jean-Pierre		20th Century Fox	Ripley is cloned so that scientists can take the Queen Alien out of her. Ripley's DNA gets mixed up with the Queen's and she begins to develop certain alien characteristics. The scientists begin breeding the aliens, but they later escape. The Queen then gives birth to a new breed of human/alien.	clone, genome	horror, science fiction

Films *(continued)*

Title	*Date*	*Director*	*Country*	*Studio*	*Link to genomics*	*Keywords*	*Genre*
Doppelganger	1997	Siskin, Nick	UK/ Australia	Digital Pictures/ Passion Pictures	City boss Reggie Backhander has been creating experimental clones of great British footballers, but adding ape DNA to increase their strength and aggression.	clone, DNA	animation short
Gattaca	1997	Andrew, Niccol	USA	Columbia Pictures	Vincent is one of the last 'natural' babies born into a genetically enhanced world, where life expectancy and disease likelihood are ascertained at birth. He has no chance of a career in a society that now discriminates against your genes.	genetic selection	science fiction
Lebensborn	1997	Stephens, David	USA	Remington York Inc.	Two freshmen unknowingly become part of a scheme to create superior humans through genetic engineering.	gene, genetic	horror
Lost World: Jurassic Park	1997	Spielberg, Steven	USA	Universal Pictures/ Amblin	Sequel – a scientific team is sent to a second island populated by cloned dinosaurs, which appear to have established themselves.	clone	science fiction, action
Mimic	1997	del Toro, Guilliame	USA	Miramax	Scientist genetically engineers a new insect (the Judas Breed) from termite and mantid DNA, which kills the cockroaches that spread a deadly disease. However, the modified bug mutates and mimics its new prey – humans.	gene, genetic, DNA, mutant	horror, science fiction

Morella	1997	Dudelson, James Glenn	USA	Allott Productions	Cloning and genetics experiments going wrong.	gene, genetic, clone	science ficton, thriller
Parasite Eve	1997	Ochiai, Masayuki	Japan	Toho Company Limited	Toshiaki Nagashima is a biologist who is doing major research on mitochondria. When his beautiful young wife is tragically involved in a car accident which leaves her brain-dead, in desperation he steals her liver in order to retrieve the mitochondria from it to resurrect her from the dead. The killer mitochondria takes the form of his assistant who then uses the biologist as a host for a terrifying new species that threatens to take over the world.	DNA	horror, science fiction
Alone: Life Wastes Andy Hardy	1998	Arnold, Martin	Austria	SixPack Film	Andy Hardy, the all-American sunny boy of the 1930s and 1940s, returns as an oedipally destroyed teenie clone to be released from his suffering by Betsy's singing and kisses.	clone	musical, short
Blue Submarine (Ao no roku gô)	1998	Maeda, Mahiro	Japan	Bandai Entertainment Inc.	Based on the premise of Noah's Ark and *The Island of Doctor Moreau*, the story revolves around a doctor's revenge on humanity by creating mutants that eventually wage war against humans.	mutant	horror, science fiction

Films *(continued)*

Title	*Date*	*Director*	*Country*	*Studio*	*Link to genomics*	*Keywords*	*Genre*
Prince of Darkness (*Kido Senkan Nadeshiko*)	1998	Sato, Tatsuo	Japan	Xebec	Fourteen-year-old genetically engineered wunderkind Ruri Hoshino is now captain of the new battleship *Nadesico*.	gene, genetic	action
Soldier	1998	Anderson, Paul W. S.	UK/USA	Impact Pictures	Breeding genetically enhanced soldiers.	genetic	action, science fiction
Species II	1998	Medak, Peter	USA	MGM	Astronaut Patrick Ross is infected with alien DNA on a trip to Mars. Back on Earth he spreads it through sex, to create instant human-alien hybrid children. A multi-disciplinary team tries to find him, with the aid of Eve – a clone from Sil's original DNA. Her alien genes are suppressed, but can her humanity survive alien sex?	DNA, clone, hybrid	science fiction, thriller, horror
After the Truth	1999	Richter, Roland Suso	Germany	Helkon Media	Josef Mengele is bought to trial in modern-day Germany to account for his part in genocide and his horrific medical research done in the name of eugenics.	eugenic	thriller
Austin Powers: The Spy Who Shagged Me	1999	Roach, Jay	USA	Eric's Boy	Clone of Dr Strangelove spoof.	clone	comedy

Deep Blue Sea	1999	Harlin, Renney	USA	Warner Brothers	Sharks are genetically modified in order to obtain a treatment for Alzheimer's disease.	gene, genetic	horror
eXistenZ	1999	Cronenberg, David	Canada/UK	Screen Ventures	Genetically engineered amphibians and reptiles are made into organic games modules in lab/abattoir/factory complex. The creatures also turn up on the menu of a nearby restaurant – GM food and novel taste sensations.	gene, genetic, GM food	science fiction
2001: A Space Travesty	2000	Goldstein, Allen	Germany	Helkon Media	President of the United States is being held captive on a secret international moon base called Vegan and has been replaced on Earth by a clone.	clone	comedy, science fiction
6th Day, The	2000	Spottiswoode, Roger	USA	Phoenix Pictures	Futuristic action about a man who meets a clone of himself and stumbles into a grand conspiracy about clones taking over the world.	clone	action, science fiction, thriller
Hollow Man, The	2000	Verhoeven, Paul	USA	Columbia Pictures	Explores the repercussions of a genetic scientist's search for the formula for invisibility. The protagonist is portrayed as an archetypal 'mad scientist' who becomes dangerously focused on his research. He successfully makes animals invisible but then insists on becoming the next test subject. The scientist becomes invisible but slowly becomes insane and begins to stalk the members of his research team.	gene, genetic, scientist, serum	science fiction, horror

Films *(continued)*

Title	*Date*	*Director*	*Country*	*Studio*	*Link to genomics*	*Keywords*	*Genre*
Mission to Mars	2000	de Palma, Brian	USA	Touchstone Pictures	DNA turns out to be left by aliens from Mars, later used as a form of signalling, explaining the origin of human life on earth and creating kinship with extra terrestrials.	DNA, genetic manipulation	action, thriller
Nutty Professor II: The Clumps	2000	Segal, Peter	USA	Imagine Entertainment	This sequel to the Nutty Professor is a slapstick take on Dr Jekyll and Mr Hyde.	DNA, serum	science fiction, comedy, romance
X-Men	2000	Singer, Bryan	USA	20th Century Fox	Set in the near future, where children are being born with a special X-Factor in their genes, giving them special powers, and making them 'mutants'.	gene, mutant	action, science fiction, thriller
American Astronaut	2001	McAbee, Cory	USA	BNS Productions	Curtis receives a home-made cloning device already in the process of creating a creature most rare in this space quadrant – a Real Live Girl.	clone	comedy, musical, science fiction
Doppelganger	2001	Horowitz, Michael and Smith, Gareth	USA	Not available	When Brian's girlfriend stays awake to watch him sleep, she finds that he has a clone who wants to take over.	clone	short, thriller

Black Mask 2: City of Masks	2002	Tsui, Hark	Hong Kong	China Star Entertainment	Genetic modification means Black Mask has superhuman fighting and healing powers.	gene, genetic, mutant	action
Blade II	2002	del Toro, Guilliame	USA	Amen Ra Films	Genetically modified vampires infect humans and vampires.	hybrid	horror, science fiction
Die Another Day	2002	Tamahori, Lee	UK/USA	MGM	DNA transplanting is used to change identity and increase longevity for the very rich.	DNA	action, Bond
Made Incorrect	2002	Lindquist, Aaron	USA	Aristaeus Productions	Mother miscarries and then grows clone in test tube.	clone	horror, science fiction, short
Nightingale in a Music Box	2002	McDermott, Hurt	USA	Rebus Productions	Patenting of new life forms which can change human memory (nanobots) whose genetic blueprints are then stolen.	gene, genetic	thriller
Resident Evil	2002	Anderson, Paul W. S.	UK/Germany/ Pathé/France/ USA	Constantin Film Produktion	A special military unit fights a powerful, out-of-control supercomputer and hundreds of scientists who have mutated into flesh-eating creatures after a laboratory accident.	mutant	horor, science fiction, action, thriller
Spiderman	2002	Raimi, Sam	USA	Columbia Pictures/ Marvel	Body chemistry is mutagenically altered in that he can scale walls and ceilings.	genetic, mutant	action, science fiction

Films *(continued)*

Title	*Date*	*Director*	*Country*	*Studio*	*Link to genomics*	*Keywords*	*Genre*
Star Wars II: Attack of the Clones	2002	Lucas, George	USA	20th Century Fox	Clone army grown for the Jedis gets co-opted by the evil empire and develops into the storm troopers.	clone	science fiction, action
Teknolust	2002	Hershman-Leeson, Lynn	USA	Blue Turtle	Rosetta Stone (Tilda Swinton), a biogeneticist creates a recipe for cyborgs and uses her own DNA in order to breed three self-replicating automatons, part human/part computer.	DNA, clone	drama, science fiction, romance
Tuxedo, The	2002	Donovan, Kevin	USA	Dreamworks	Disease developed to contaminate the water supply.	genetic	comedy
Yesterday	2002	Youn Soo, Jung	South Korea	Miracin Korea Film Company	Genetic manipulation of DNA and chromosomes in order to alter behaviour and agression.	DNA, gene, genetic	action, science fiction, thriller
Agent Cody Banks	2003	Zwart, Harald	USA/ Canada	MGM	A scientist unknowingly develops a fleet of deadly nanobots for the evil organisation ERIS.	genetic	action
Blood of the Beast	2003	Koszulinski, Georg	USA	Substream Films	The first strand of clones are harvested in December of 2012 and received with overwhelming success. It is not until nineteen years later that the first problems arise.	clone	action, science fiction

Blueprint	2003	Schubel, Rolf	Germany	Relevant Film GmbH	Iris Sellin learns that she has an incurable illness. In order to preserve her art and also herself, beyond death, for all posterity, she has herself cloned. Daughter Siri is her virtual twin and learns as a child that she is the world's first cloned human being.	clone	drama
Code 46	2003	Winterbottom, Michael	UK	BBC/ Revolution Pictures	Futuristic drama/love story about genetic incompatability through shared DNA. In the futuristic setting, IV methods and so on have reached a point where genetic testing is required before sex to check that too much is not shared.	gene, genetic	drama, science fiction, romance
Hulk, The	2003	Lee, Ang	USA	Universal Pictures	Geneticist's experimental accident curses him with the tendency to become a powerful giant green brute under emotional stress.	gene, genetic, monster	action, adventure, science fiction
If	2003	Stoll, Lisa	USA	Space Dawg Productions	Science and religion collide when Josh finds out that he was cloned.	clone	science fiction, thriller
Numb	2003	Gibson, Michael Ferris	USA	Yerba Buena Productions	Pharmacogenomics and genetic engineering.	gene, genetic, genomic	science fiction
Save The Green Planet	2003	Seung-Jae, Cha	South Korea	Sidus	Alien film with some discussion of human genetics.	n/a	thriller/ black comedy/ science fiction

Films *(continued)*

Title	*Date*	*Director*	*Country*	*Studio*	*Link to genomics*	*Keywords*	*Genre*
X-Men 2	2003	Singer, Bryan	USA	20th Century Fox	More mutants.	gene, mutant	action, science fiction, thriller
Able Edwards	2004	Robertson, Graham	USA	Graphic Films	About the clone of a famous entertainment mogul created to revive the glory days of his deceased predecessor's corporation. In the process of restoring reality entertainment to a synthetic, virtual world, the clone realises he has yet to live as his own man.	clone	drama, science fiction
Casshern	2004	Kiriya, Kazuaki	Japan	Casshern Film Partners	The geneticist Dr Azuma vies for support from the government for his 'neo-cell' treatment that he claims can rejuvenate the body and regenerate humankind. Mutants unleashed.	gene, genetic, mutant	action, adventure, science fiction
Godsend	2004	Hamm, Nick	USA	2929 Productions/ Artists Production Group	After their son, Adam, is killed in an accident, a couple approach an expert in stem-cell research about bringing him back to life through an experimental and illegal cloning and regeneration process.	clone, stem cell	drama, fantasy, horror, thriller
Manchurian Candidate, The	2004	Demme, Jonathan	USA	Paramount Pictures	In the midst of the Gulf War, soldiers are kidnapped and brainwashed for sinister purposes.	biotech-nology	science fiction, thriller

Resident Evil: Apocalypse	2004	Witt, Alexander	UK/Germany/ France/USA	Columbia	More of the same, different location. (See *Resident Evil*, 2002.)	mutant	horror, science fiction, action, thriller
Ultraviolet	2004	Wimmer, Kurt	USA	Ultraviproductions Inc.	A subculture of humans who have been modified genetically by a vampire-like disease are seen as a threat.	gene, genetic	action, science fiction, thriller
Aeon Flux	2005	Kusama, Karyn	USA	MTV/Paramount	Four hundred years in the future, when disease has wiped out the majority of the earth's population except for one walled, protected city-state: Bregna, ruled by a congress of cloning scientists.	clone	science fiction, action
Cl.One	2005	Tomaric, Jason	USA	Bell tower/Image Entertainment	Humanity is on the verge of extinction and only cloning can save it.	clone	horror, science fiction
Island, The	2005	Bay, Michael	USA	Dreamworks	Lincoln Six-Echo and all of the other inhabitants of the facility are actually human clones whose only purpose is to provide 'spare parts'. Lincoln escapes with a beautiful fellow resident named Jordan Two-Delta.	clone, organ	drama, science fiction
X Men III	2006	Ratner, Brett	USA	Twentieth Century Fox	A 'cure' for mutation is found.	mutation, genetic engineering	science fiction, action

Television documentary

Title	*Date*	*Country*	*Company*	*Plot*	*Keywords*
Life Story	1987	UK	BBC	Francis Crick and Jim Watson in Cambridge lead the race to discover the structure of DNA, with the aid of Maurice Wilkins in London, and in competition against Rosalind Franklin in Paris.	DNA
Synthetic Pleasures	1995	USA	Caiparihnia Productions	Investigates cutting-edge technologies and their influence on our culture as we approach the twenty-first century. A new world is suddenly emerging, an artificial reality. Virtual reality, digital and biotechnology, plastic surgery and mood-altering drugs promise seemingly unlimited powers to our bodies and ourselves. This film presents the implications of having access to such power as we all scramble to inhabit our latest science fictions.	biotech
Frontline Scotland: Cloning and Parkinson's Disease	1999	UK	BBC Scotland	Explores controversies in science including potential use of embryos.	clone, stem cell
Faces of Islam	2000	UK	BBC	M. Kabir Banu Al Zubair – a geneticist at Cambridge discusses the relationships between Islam and genetic science, especially cloning.	genetic, clone
Horizon: Life and Death in the 21st Century: Living Forever	2000	UK	BBC	Geneticists addressing the ageing process through stem cells and Progeria	genetic, stem cell

Tinok Lafi Huzmana (*Baby to Order*)	2000	Israel	Israel Broadcasting Authority	Examines the one subject that will influence our lives more and more in the near future – cracking the human genome and its profound consequences. The next generation can be born perfect, genetically speaking. It is possible to envision a world without disease or without slight flaws such as baldness or the need for glasses. What are the advantages and disadvantages of such a situation? What constitutes the red line beyond which it is no longer worth being born?	genome
Horizon: The A6 Murder	2001	UK	BBC	Can DNA solve the question of the identity of the killer?	DNA
Horizon: Cloning the First Human	2001	UK	BBC	Is the science actually ready yet for cloning healthy humans? Horizon follows the latest research, which has led many scientists to believe that Zavos and Antinori's plans to clone the first human could end in tragedy.	clone, genetic
Horizon: Dawn of the Clone Age	2001	UK	BBC	A detailed account of the birth of Dolly and previous attempts on tadpoles and cows.	clone
Cousins	2002	UK	BBC	Observation of human/primates similarity based on the premise that they share 98 per cent of our DNA.	DNA, gene, genetic
Genetics Debate, The	2002	UK	BBC	Debate exploring key scientific and ethical issues associated with this major area of scientific advancement as well as related areas, in particular cloning and the use of stem cells for research purposes.	clone, gene, genetics, stem cell
How to Build a Human	2002	UK	BBC	Looks at the origins of life and how cloned organs might play a role in repairing damage to the human body.	clone

Television documentary *(continued)*

Title	*Date*	*Country*	*Company*	*Plot*	*Keywords*
The First Human Clone: Conception	2003	UK	Channel 4	Peter Williams follows Dr Zavos and his team and films the production of the ten-cell embryo.	clone
Of Apes and Men	2002	UK	BBC	Looking at apes as our genetic cousins.	gene, genetic
Bitter Inheritance	2004	UK	BBC	Traces the stories of the lives of families with hereditary genetic conditions.	gene, genetic
DNA's Dark Lady: Rosalind Franklin	2004	UK	BBC	Profile of Rosalind Franklin and DNA 'discovery'.	DNA, gene
Double Helix: The DNA Years	2004	UK	BBC	This two-part series tells the story of a fifty-year-old quest to discover the secret of human nature hidden in our DNA.	DNA, gene, genetics
Human Mutants	2004	UK	Channel 4	Armand Leroi hosts exploration of human genetics through mutation.	mutation, clone
Richard Dawkins: Profile	2004	UK	BBC	Profile of the pioneering evolutionary biologist and author of *The Selfish Gene.*	gene
Superfly	2004	UK	BBC	A fly's eye view of 100 years of genetics. *Drosophila melanogaster* – the common fruit fly – might look insignificant but over the last century it has been used to unlock the secrets of life. Sixty per cent of our genes are the same as the fly's.	gene, genetic
Trial and Error	2004	UK	BBC	Gene therapy experiment goes wrong.	gene

PD Notebook	2005	South Korea	Munhwa Broadcasting Corporation	South Korean investigative documentary with episodes in 2005 showing that there were problems with how Hwang obtained women's eggs, and presenting evidence of fraud.	n/a
What Makes Us Human	2006	UK	Channel 4	Armand Leroi hosts exploration of what makes us human via genetics.	genetics

Television drama

Title	*Date*	*Country*	*Production Company*	*Plot*	*Key word*	*Genre*
Dr Who	1963 – onwards	UK	BBC	Time lord biology allows regeneration of identity in new body.	regeneration	adventure, drama, science fiction
Star Trek	1966 – onwards	USA	CBS/Paramount	Some cloning episodes.	space	science fiction
Cloning of Joanna May, The	1991	UK	Granada TV	A millionaire produces three clones of his former lover.	clone	comedy, drama
Micronots, The	1993	USA		In the very distant future, humans have lost the ability to continue procreation when the world's cloning centres are sabotaged.	clone, gene, genetic	adventure, comedy, science fiction
X-Files, The	1993–2002	USA	20th Century Fox	Some cloning episodes.	science fiction	science fiction

Television drama *(continued)*

Title	*Date*	*Country*	*Production Company*	*Plot*	*Key word*	*Genre*
Generation X	1996	USA	MT2 Services/ Marvel	The new teenage students at a school for mutants (normal-looking humans with fantastic powers) – X-men variant.	gene, genetic, mutant	action, adventure, science fiction
Crusade, The	1999	USA	Babylonian Productions	Genetic plague.	gene, genetic	action, drama, science fiction
Now and Again	1999	USA	CBS Productions	Brain transplanted into genetically engineered body.	gene, genetic	action, science fiction
Cybersix	1999	Canada/ Japan/ USA	NAO	Cybersix is a powerful advanced female genetic construct android.	gene, genetic	animation, action, fantasy
Dark Angel	2000	USA	20th Century Fox	Twelve genetically enhanced children escape from the military base that created them.	gene, genetic	action, adventure, science fiction
Learning to Love the Grey	2000	UK	BBC	Explores ethical and scientific issues around cloning through artist, scientist and patient interests.	cloning, stem cells	drama (didactic)
Toshinden Subaru: The New Generation	2000	Hong Kong	Mandarin Films	Eiji started a team of genetically engineered beings to help him with his idea of becoming an ultimate god.	gene, genetic	animation, action, fantasy

Fearless	2004	USA	Jerry Bruckheimer	Born with genetic defect.	gene, genetic	action, drama
If . . .	2004	UK	BBC Two	Second series episode *If . . . Cloning Could Cure Us* fictional scenario – real debate.	cloning, stem cells	court room docudrama
The Eleventh Hour	2006	UK	Granada Television	First episode *Resurrection* is about human reproductive cloning.	clone, science	science fiction, drama
The Family Man	2006	UK	BBC	IVF clinic and spin-off stories including PGD and sex selection.	infertility, embryo	medical drama

Notes

1 Introduction

1 In addition to the research of Jackie Stacey (2003, 2005 forthcoming) mentioned previously, we refer here to the work of Debbora Battaglia (2001), Matt Hills (2004), David Kirby (2003a, 2003b), Brigitte Nerlich (2001), Judith Roof (1996), Megan Stern (2004), Eugene Thacker (2002, 2005), Peter Weingart et al. (2003) and Aylish Wood (2002). Christopher Frayling's (2005) study of the representation of scientists in film was another useful resource. We have also been stimulated by Susan Anker and Dorothy Nelkin's (2003) charting of recent genetic art.

2 The Mass Observation Archive was established in 1937 to encourage and collect observations of daily life in Britain and it operated into the 1950s. It was re-established, with some modifications of its mission and procedures, with its base at Sussex University in 1981. See www.massobs.org.uk.

2 What is cloning?

1 The shift from the terminology of the 'clone' to that of 'cloning' is implicit in our account. Lee Silver suggests that the terminological shift from 'clone' to 'cloning', as well as the 'fantastical prediction that "man will be able to make biological carbon copies of himself" derived from Alvin Toffler's 1970 book *Future Shock*. He asserts: 'In one fell swoop, clones morphed from the simple progeny of asexual reproduction to sophisticated products of biological engineering created by scientists bent on controlling nature' (Silver 2001). 'What are clones? They're not what you think they are' *Nature* 412 (5 July, 2001: 21).

2 Wilmut and Highfield remark that: 'The achievement of her [Dolly's] birth rested on a sturdy foundation of skills, understanding and invention that had gradually been built since in [*sic*] the nineteenth century by scientists around the world. And it relied on a zoo of species' (Wilmut and Highfield 2006).

3 *After Dolly* and its key arguments have been reviewed in local, regional and national newspapers in the UK, the USA, Canada and Australia, as well as in specialist journals such as *The Lancet*. Additional English language reviews were distributed by news wire services including the Xinhua General News Service and the state news agency of China. Ian Wilmut was interviewed about the book on National Public Radio in the USA. In Roslin's home city of Edinburgh, Wilmut and Highfield were a major draw at the Edinburgh Book Festival in August 2006.

4 At a workshop at the Max Planck Institute in Berlin in March 2007, Manfred

Laubichler, of the School of Life Sciences at Arizona State University, pointed out that Spemann's fantastical experiment had already been conceptualised by Theodor Boveri decades earlier. This historical perspective on the fields of genetics and embryology underlines the observation that there are multiple historical narratives that might be composed when (re)constructing the telos of a scientific field or object.

5 Christina Brandt claims that the first embryologists to use the term 'clone' were Robert Briggs and Thomas King in their paper 'Serial transplantation of embryonic nuclei' (Briggs and King 1956).

6 Illmensee is credited as joint author with Panos Zavos on a paper about human reproductive cloning published in *Archives of Andrology* (Zavos and Illmensee 2006).

7 In a review of *After Dolly* Steven Rose points up the irony of this approach in 'an unashamed work of advocacy' for therapeutic cloning. Rose wonders: 'wouldn't animal studies be the more appropriate and ethical place for research to start?' (Rose 2006).

8 The dedication in *Solution Three*, Naomi Mitchison's (1975) science fiction novel, reads: 'To Jim Watson, who first suggested this horrid idea.' So the direction of travel between the imaginations of scientists and the pages of science fiction novels is never simply one-way.

9 According to their website (www.thehastingscenter.org): The Hastings Center is an independent, non-partisan, and non-profit bioethics research institute, founded in 1969 to explore fundamental and emerging questions in health care, biotechnology and the environment.

10 Peter Poon also notes that: 'Cloning . . . became one of the first issues tackled by the emerging discipline of bioethics in the late 1960s and early 70s' singling out two theologians, Paul Ramsey and Joseph Fletcher, as exemplars. He points out that they adopted opposing ethical positions on the subject of human cloning (Poon 2000: 165).

11 Bromhall notably co-operated in an iconic fictional portrayal of cloning. He was credited as science adviser on the 1978 film *The Boys from Brazil*.

12 Poon goes on to point out that, in 1979, Landrum B. Shettles reported in the *American Journal of Obstetrics and Gynecology* that he had produced three human embryos using what we would now call somatic cell nuclear transfer. He claimed that he had then destroyed the embryos. Through a combination of lack of supporting evidence and scepticism because of his prior involvement in research-related controversy, Shettles' claims have slipped into obscurity (Poon 2000). It is notable that Shettles does not appear in the roll call of maverick cloners identified in this chapter and does not even have the name recognition of David Rorvik.

13 Parrinder (1980) and Suvin (1979) provide scholarly overviews of science-fiction literature. Armitt (1991), Barr (1993) and Donawerth (1997) focus particularly on women and feminists writing science fiction. Kuhn (1990) and Sobchack (1997) provide analyses of science-fiction film. Haraway (1991) and Balsamo (1996) examine the intersections between technoscience and science fiction.

14 We want to resist placing therapeutic cloning on the opposite pole of a potential/actual opposition as it remains to be demonstrated that experiments with somatic cell nuclear transfer on human oocytes have therapeutic application.

15 In 1993, Jerry Hall and Roger Stillman's report that they had cloned human embryos through inducing IVF embryos to split led to a flurry of public anxiety that could be understood as a dress rehearsal for post-Dolly concerns (Poon 2000).

16 Sarah Franklin suggests that this image signifies 'the global biological' (Franklin 2005).

17 According to *The Cloning Sourcebook* (2003), 'Arlene Judith Klotzko, M. Phil., J.D., is Writer in Residence at the Science Museum, London, Advisor on Science and Society to the MRC Clinical Sciences Centre, and Visiting Scholar in Bioethics at the Windeyer Institute, University College, London, England'. According to the website of the Gordon Poole Agency, she is a bioethicist and lawyer, consultant of and writer on biotechnology (www.gordonpoole.com/?ArtistID=275 – consulted 23 February 2007).

18 See Holliman for a nuanced account of how media coverage in the UK 'began by reporting the scientific announcement, but was then sustained by an emerging political and ethical controversy, mainly involving politicians, religious figures and scientists' (2004: 115).

19 The 'baby without blemish' story is the formulation Mulkay uses to encapsulate the narrative structure of press stories that focused on textual and visual representations of the healthy children born through IVF to emphasise the benefits of embryo research.

20 In January 2004, when Zavos announced that this was a service he was prepared to make available to clients from the UK, the HFEA issued a press statement reiterating that clinics in the UK were expected not to split embryos for use in treatment. See: www.hfea.gov.uk/cps/rde/xchg/SID-3F57D79B5BD19FC6/hfea/hs.xsl/ 1031.html

21 'The Corner House is a research and solidarity group which aims to support the growth of a democratic, equitable and non-discriminatory civil society in which communities have control over the resources which affect their lives and livelihoods, as well as the power to define themselves rather than be defined only by others' (Sexton 1999).

22 See Jackie Stacey's argument in 'She is not herself: the deviant relations of *Alien Resurrection*' (Stacey 2003) on the difficulty of representing genetics in the visual register.

23 He has, however, applied to the HFEA for a licence to conduct SCNT research using animal eggs that would result in 'hybrid embryos'.

24 The establishment of the Science Media Centre in 2002 can be taken as indexical of the concerns that UK scientists have with the media as a communication channel for transmitting the right message about science. The Science Media Centre describes its vision as: 'Good public policy decisions on science based on a more balanced, rational, accurate debate within the news media about science issues' (Science Media Centre 2002). It reports that its creation was in response to the House of Lords Select Committee on Science and Technology's Third Report (2000) on 'Science and Society', which suggested that a new culture of open and positive communication with the media was required in order to renew public trust in science.

25 Panos Zavos has published on his clinical research in the *Archives of Andrology*, but only the *Guardian* in the UK even noted this, dismissing it as an obscure journal.

26 PROGRESS was formed in 1985, initially with the longer title 'PROGRESS: Campaign for Research into Reproduction' and the group's first act was to launch a campaign to increase public understanding of and support for embryo research. (Although Mulkay renders the organisation's name as Progess, it was originally

produced in all upper case characters.) Professor Robert Winston was the organisation's president. (Mulkay 1997). The organisation achieved its aim, the passage of the Human Fertilisation and Embryology Act in 1990, but the Progress Educational Trust, a small charity established in 1992, characterises itself as carrying on the educational work of PROGRESS (www.progress.org.uk/About/history.html).

27 The Center for Genetics and Society, established 2001, is based in Oakland, California. It describes itself as: 'a nonprofit information and public affairs organization working to encourage responsible uses and effective societal governance of the new human genetic and reproductive technologies'. See www.genetics-and-society.org/about/index.html.

28 As a representative from the People's Solidarity for Participatory Democracy commented: 'Around the time after the second announcement [the breakthrough announced May 2005] the social mood was so serious that I felt as if I may be attacked due to my activity, because Professor Hwang had become a hero to Koreans. I mean not just cyber terror or protest by telephone but physical terror like being pelted with stones' (representative interviewed by Choon Key Chekar).

29 With the discrediting of the human embryonic cloning reported in Hwang et al. (2004) and Hwang et al. (2005) plus controversy over premature publication of results by the Newcastle Centre for Life, the only centre in the UK actively engaged in SCNT work with human oocytes, it seeems unreasonable to make emphatic claims about shifts from the imminent to the realised.

3 Cloning futures

1 The 'uncanny' is a trope of Freudian theory (*unheimlich*), where it is figured through the double – shadows, reflections, automata – and death. The horror of the uncanny is thought to be related to its familiarity, a familiarity which exceeds definition, or the experience of both knowing and unknowing at the same time. Mechanical dolls, for example, exemplified the uncanny because of their likeness to human form and their ambiguity. Twins and thus clones (as twins) are also associated with this trope.

The 'yuk factor' refers to responses or feelings in a register of disgust or repugnance. It is used to refer to a range of responses that do not lend themselves to articulation. This has been theorised in a number of ways, one of the most controversial is the theory that the 'yuk factor' has moral value and might be used as a principle for governance. The fact that 'yuk factor' levels differ, and that sexism, racism and homophobia are frequently justified through this rubric, have contributed to the overt rejection of this 'factor' in relation to governance.

2 Although films do not lend themselves to straightforwardly 'either/or' readings, earlier films locate cloning within horror and science fiction. However, since the late 1990s, films have increasingly represented cloning in mainstream genres and the association with horror seems to be decreasing. These five films all represent cloning as naturalised, as a potential reproductive technology and as a potential health care solution.

3 We used the Mass Observation Archive at the University of Sussex to send out a directive, in spring 2006, asking respondents to write about how they knew about genes, genetics and cloning. We directed them to consider as wide a range of sources as possible and we received 174 responses. The questions in the Mass Observation

Archive directive were composed by the archive staff and the authors of this book. The responses were individual written replies to the directive from regular archive respondents. All this material was collected in 2006 by the project team.

4 In addition to the Mass Observation Archive respondents, the participants referred to here include face-to-face focus group meetings with a researcher, which were recorded. We also conducted workshops with school groups, and organised discussions after film screenings with other groups. We conducted individual interviews with stem-cell and cloning scientists, and members of key non-governmental organisations, which we draw on here.

5 Rather than indicating lay ignorance, these comments are indicative of sensitive readings of heightened expectations regarding cures and regeneration through cloning (Irwin and Wynne, 1996).

6 Harry Griffin, the CEO of the Roslin Institute, has a different take on this. He sees the media leaping to conclusions about reproductive cloning based on an over-arching narrative that 'science is moving too fast': 'The news media are obsessed by timeframe', he told us, 'So the fact that Dolly was already seven months old or whatever, before she was [revealed in the media] obviously meant that we had several women pregnant. That wasn't said specifically, but that was the general flavour. Somebody from the HFEA had been asked how long would it take to get a licence to clone human embryos and she said, "Well, it'll take about – if we receive an application we would hope to process it within a year." So: "cloned human embryos within a year". So speed and the general theme is that science is moving so fast that we can't regulate, we're going to be overwhelmed by the issue. Well, it's not happened, has it?' (Harry Griffin, interview with the authors).

7 LIFE – 'the UK's leading pro-life charity' www.lifeuk.org

8 This report is published by the UK House of Commons; www.publications.parliament.uk/pa/cm200405/cmselect/cmsctech/7/702.htm

9 Ian Wilmut was cited in the report from a *New Scientist* article 21 February 2004; Sarah Parry appeared as a witness and gave oral evidence for this report www.publications.parliament.uk/pa/cm200304/cmselect/cmsctech/uc599-ix/uc59902.htm

10 We ran two workshops at the University of Lancaster and a workshop in a local school for Social Science Week, in 2005 and 2006. The 2006 workshop was attended by Year 11 and Year 12 groups from two local schools and involved discussion and debate followed by students creating their own cloning headlines; www.cesagen.lancs.ac.uk/events/past/socialscienceweek2006.htm www.cesagen.lancs.ac.uk/events/past/socialscienceweek2005.htm

11 The Human Cloning film festival was organised by the Scottish Council for Bioethics, the South-East Scotland Branch of the British Association for the Advancement of Science, and The Edinburgh Film House. It was funded by the Wellcome Trust, the Genomics Forum and Awards for All Scotland. The festival attracted 300 participants and showed fifteen films including science communication, documentary and Hollywood film.

12 Pseudonyms have been used.

13 The Wellcome Trust is a UK biomedical research trust. Whilst it has the status of a charity, it is also the largest source of non-governmental funding for biomedical research in the UK. It has an annual spend of around £500 million, which goes to support work both in the UK and internationally.

14 This paper was based on an analysis of UK press and TV reporting on embryonic

stem-cell research. It examined the different rhetorical strategies used to constitute different ethical positions.

15 Comedy does not fit into the temporal arc of other films, where cloning has been moving from a negative to a positive trope. This genre renders cloning laughable and undercuts its scientific possibility. Thus films such as *Multiplicity* (1996) and *Austin Powers: The Spy Who Shagged Me* (1999) represent cloning as an intertextual filmic motif, not as a near possible scientific practice.

4 Mavericks, madmen and fallen heroes

1 In the 1976 science fiction novel *Cloned Lives*, the first human clones born are produced only after the expiration of a moratorium on genetic research that evokes echoes of the early 1970s concerns voiced by genetic researchers and discussed at the 1975 Asilomar Conference. This extract resonates with van Dijck's (1998) analysis:

> Paul remembered the arguments made by those who had desired the moratorium. An analogy had been drawn between the biological sciences and nuclear physics and a question posed: why wait until the biological equivalent of an atomic bomb was developed before doing something? Why not prevent its occurrence? Biology presented a threat to human society and evolution far greater than that of atomic weapons. It might enslave people or alter them beyond recognition. If used foolishly, biological engineering might set humanity on an evolutionary path leading it to extinction.
>
> (Sargent 1976)

(cf. Turney (1998); van Dijck (1998).)

2 Other scholars have explored Newton's changing significance in British and European culture. See, for example: Fara (2002); McNeil (1988).

3 Arlene Klotzko identifies Zavos as one of 'a quartet of would-be human cloners' that includes Richard Seed, Rael (leader of the Raelian cult) and Severino Antinori, 'the rather operatic Italian fertility specialist' and a former business associate of Zavos (Klotzko 2004).

4 The interview was conducted by telephone on Friday, 9 March 2007.

5 Zavos specified England rather than the UK or Great Britain. We did not probe on this, but it is notable that on the website of the Zavos Organization (discussed further in Chapter 6) an open letter that Zavos and his former collaborator, Severino Antinori, addressed to the UK press was published in the Glasgow (Scotland)-based *Sunday Herald*.

6 Mephistopheles is the name given to the devil in Goethe's version of the *Faust* story.

7 At the time of publication of the article in *The Financial Times*, Professor Moon Shin-yong ran a stem-cell research facility at Seoul National University Hospital, and was a member of the research team led by Hwang.

5 Women's bodies in cloning discourses

1 Cloning is not yet entirely normalised or sanctioned, although therapeutic cloning is approaching such normalisation in the UK. This is in contrast with the technoscientific practice of IVF which has gained notable cultural legitimisation and which, it has been argued, is increasingly naturalised in many Western countries, particularly the UK. See Throsby 2004; Thompson 2005.

2 See also Celia Roberts and Karen Throsby's work (Roberts and Throsby 2006), Sarah Parry's work on the construction of 'spare' (Parry 2006), Charis Thompson on excess (Thompson 2005) and Sarah Sexton on 'waste/surplus' (Sexton 2000).
3 See the Centre for Genetics and Society for collected news archives. The 'past news' section under their resources section includes archived coverage of these issues: www.genetics-and-society.org/news.asp
4 See for example, 'Blair courts US stem cell scientists' (Rosemary Bennett, *The Times*, 30 July 2006).
5 NESCI (2006) is a collaboration including The Institute for Stem Cell Biology and Regenerative Medicine at Durham University, Newcastle University, Newcastle Centre for Life, the National Health Service (NHS) and the regional development agency One NorthEast.
6 HFEA consultation document: 'Donating Eggs for Research: Safeguarding Donors' www.hfea.gov.uk/cps/rde/xbcr/SID-3F57D79B-7764463D/hfea/donating_eggs_for_research_safeguarding_donors_consultation_FINAL.pdf
7 http://2006forum.womenlink.or.kr/intro.php
8 See Chapter 4 for further analysis of this reporting.
9 We explore an example of this below, but this occurred over multiple sites. Naked women featured prominently in the UK press coverage the joint Newcastle and South Korean announcements in May 2005 and the *Mirror*'s health section (7 May 2005) used a full body image of a naked women to illustrate a feature story about stem-cell cures based on the May announcements (UK).
10 See Chapter 3 and also organisations such as the Christopher Reeve Foundation, Michael J. Fox Foundation, Motor Neurone Disease Association.
11 This comment was one of a collection solicited by the Science Media Centre in the UK as a response from the scientific community to Ian Wilmut's 2004 suggestion that women should donate eggs for therapeutic cloning research. Science Media Centre press release 26 July 2005, 'Scientists React To News That Ian Wilmut Is Seeking HFEA Approval To Ask Women To Donate Eggs For Research'. http://www.sciencemediacentre.org/press_releases/05-07-26_wilmutegg donors.htm
12 As is also documented by researchers in fertility clinics (Crowe 1990; Thompson 2004; Throsby 2005) most women recognise that they must self-manage their image to get access to IVF. Most of them recognise that they should appear as simultaneously desperate enough, but not too desperate.
13 The full text and full list of signatories to this letter can be found on the Science Media Centre website at www.sciencemediacentre.org/press_releases/04-01-21_Zavos_open_letter.htm
14 The negative impact of being a clone is explored in the films *Blueprint* and *Godsend*.
15 This name is a play on several different levels. The Rosetta Stone was a historical artefact, the discovery of which allowed translation between two ancient languages. It has become symbolic of 'translation' more generally. The link between this and Sandy Stone – transgender, cybertheorist and performance artist – provides another play across cyberfeminism (there is another character called Sandy in the film). Rosetta Stone is also the name of an Apple programming language and Apple hardware appears throughout the film.
16 This conflation of the status of the embryo with the ethical concerns links to the way these debates have been polarised in the USA. There is also growing activism among women's health activists and those concerned with distributive justice. Leading campaigner, Judy Norsigian from Our Bodies Ourselves (also known as the

Boston Women's Health Collective), for example, describes herself as 'a supporter of most embryonic stem cell research' but with substantial concerns about the use of SCNR embryos. She has been a prominent figure in the call for a moratorium on such research in the USA. Although she is often cast by the media as a 'strange bedfellow' with the Catholic Church, this is a positioning she vehemently rejects. 'This is a colossal myth that it's a matter of the pro-choice, pro-science votes on the one side and the religious social conservatives on the other side', she told us. There are, she points out, legitimate health and safety reasons for objecting to some forms of stem-cell research, and these reasons are generally not the primary concerns of the anti-abortion lobby, which focuses on quite different distinctions. 'Although there are those who have deliberately confused this issue, sometimes conflating embryo cloning research with ALL embryo stem cell research, it is important to keep the two separate and to insist that health concerns for women don't take a back seat' (Norsigian, 2005: 3).

17 According to the Raelian's website, 'life on Earth is not the result of random evolution, nor the work of a supernatural "God". It is a deliberate creation, using DNA, by a scientifically advanced people who made human beings literally "in their image" – what one can call "scientific creationism."' (Rael.org)

18 For example, both James Watson and Bill Clinton claimed that DNA is the language of God (Kay 2000; Nerlich et al. 2002; van Dijck 1998).

19 The Corner House and the Center for Genetics and Society have been mentioned previously. Hands Off Our Ovaries was formed in the USA in relation to the issue of egg sourcing for cloning research as it arose in the USA and UK in 2006.

20 Focus group participants sometimes joked about how cloning might render men redundant, and male participants sometimes expressed some anxieties about the outcome of such 'queer' reproduction. One man said cloning would allow for a world dominated by lesbians, another expressed concern that cloned babies would all be female, a third said any boy child produced through cloning would probably be a 'poofter'.

21 Cloning destabilises paternity partly through the emphasis on eggs, and mainly through the move away from fertilisation. Despite media representations to the contrary, the production of SCNT embryos does not require male gametes. For a more extended analysis of anxiety about paternity in the context of new technologies see Judith Roof (1996).

22 The only female cloning scientists to appear in the news coverage were Alison Murdoch and Brigitte Boisselier. See Chapter 4 for further elaboration on the figure of the female scientist.

6 Truth claims and genres

1 We are also interested in the ways in which the mobilisation of common tropes, discursive and visual, lead observers to think of the media as 'an amorphous, interconnected, mutually dependent bunch' (Gregory and Miller 1998: 104).

2 We would suggest that Hollywood film, and TV and press news, (mis)understood as a single genre, dominate the imagined terrain of science in the media.

3 For example, in the chapter 'Rotwang and Sons' he provides a fascinating account of interactions between rocket scientists transplanted to the USA from Germany in the aftermath of World War 2, Walt Disney and the emergent US space programme.

4 One of the means by which news accounts attempt to achieve accuracy – to represent 'the truth' – is by assembling a range of sources who represent diverse of points of view. It is notable in the making of genomic science in the media – in the specific cases of cloning and stem-cell science – that balance is increasingly achieved by assembling a preponderance of speakers with epistemic authority including 'reputable' scientists and regulators. Other points of view are marginalised or reduced to in-passing comments at the end of articles. In UK press and TV news reporting some points of view, such as those held by people who are morally opposed to embryonic research, seem to be regarded as unbalanced by definition. They may be given equal airspace at times of controversy but they are not framed as 'experts' and their contributions fade from prominence once an expert and public 'consensus' has been established (Kitzinger et al. 2007).

5 Immediately before the book went to press we referred to the website again. The site's production values have been significantly improved and it features a much less cluttered visual address, making the enterprise appear more professional. Navigation through the site is also easier as the signposting has been improved. Nonetheless, the key elements of the home page remain as analysed in October 2006, and the key discursive strategies that we identified are still deployed.

6 The version of this homepage consulted on 15 March 2007 has the updated banner: 'Meet Dr Zavos, The Man That Created The First Human Cloned Embryo'. Clicking on this banner takes you to a page that reproduces the editorial from *Reproductive Biomedicine Online* in which Robert Edwards contrasts Zavos favourably with Hwang. The salient sentences are highlighted.

7 In March 2007 an additional link had been added. This link takes site visitors to a PDF version of the article published in *Archives of Andrology* in June 2006 that documents the attempt at human cloning that Zavos announced at the January 2004 press conference. The citation for the Zavos, Illmensee and Levanduski paper has also been updated to enable visitors to view a PDF version of the earlier article. (NB: PDF stands for Portable Document Format and enables any user to read an electronic document with a freely provided viewer. The PDF coding also prevents users from tampering with the documents.)

8 As already noted, between October 2006 and March 2007 many of these criticisms have been addressed. In the interview conducted with Zavos on 9 March 2007, he pointed out that he had fifteen websites and indicated that he had replaced the webmaster responsible for the October 2006 overhauls of the websites.

7 The constitution of public audiences

1 Looking at how publics are approached as research subjects is an additional way of considering how they are constructed. In this context, the contrast between how 'publics' are constructed in the US and in the UK research context is quite striking. Fiona Coyle's initial review offers systematic documentation of how the public have been researched in the USA, predominantly (and extensively) by opinion poll research (generally initiated by stakeholder groups and contracted out to specialist firms). These surveys tend to collapse 'publics' into voters whose statistical profile slots into one of two categories: 'party affiliates' or 'religious persons'. These studies are utilised to predict outcomes or even influence outcomes at times of presidential elections. In contrast, studies in the UK have been more *qualitative*, with data gathered through interviews and engagement events, with the impression being

given that these form part of consultations which might influence policy. Alongside this, questions in the USA are primarily an attempt to determine 'how' the public will vote while the questions explored in UK research are more likely to attempt to draw out the role of underlying attitudes and values in individual responses. This has interesting implications for how publics are being conceived in each country (Fiona Coyle, working paper, for 'Discourses/Publics' research).

2 See Mulkay (1997) in relation to the same question in the UK embryo debates of the 1990s.

3 A group of people with spinal cord injury who viewed *If* as part of Grace Reid's research (Reid 2007) were very aware that the appeal to the public addressed by *If* was an *able-bodied* public, appealing to their fear of becoming disabled. After watching this programme they were highly critical of the representation of the individual with paraplegia. One commented, 'I don't like the idea that the guy had no life'; another responded, 'The fact that he's a para and life's over (laughter) I mean I'd kill to be a para [. . .] I'd rather be a para than a quad. So yeah, I don't have full use of both arms, don't go and tell *me* your sad story and not do anything to deal with it.' Others in the group added comments such as, 'They made the guy look so freakin' pathetic I couldn't believe it.' These research participants differentiated themselves from an able-bodied audience commenting, 'I think as people in chairs we're going to be sceptical of that show as compared to people who aren't.' They felt that the 'man in the street' might be easily manipulated by such representations: 'If you just pull people off the street and show them that, they'd probably be "Oh my God, the poor guy. He's got no life and he's gonna die so we better get this passed so therapeutic cloning can happen because how much more terrible can life be?"' They contrasted this with their own more sceptical position: 'Yeah, we certainly don't have the same sympathy for the Andrew guy'.

4 Judaeo-Christian religions are represented as constitutive of the legitimate political subject in the current administration in the USA, whilst the category of religious extremism is reserved for non-Christian religions such as Islam.

5 Although we have not had time to review this programme in great detail, *The 11th Hour* (Granada Television, 2006) was a significant text in the context of the project. The first episode, 'Resurrection', featured a storyline about reproductive cloning. At the same time as it was launched as a fictional science drama, it had a strong didactic element and efforts were made to represent 'topical' science accurately, realistically and responsibly. The protagonist – a 'government scientist' – delivered several monologues on the importance of therapeutic cloning and the evils of reproductive cloning. This episode was purportedly inspired by Professor Winston's treatment of Dr Panos Zavos, which the scriptwriter – Steven Gallagher – interpreted positively. Steven Gallagher also said that he had deliberately tried to attempt a responsible representation of science in the drama.

6 The focus groups were all face-to-face group meetings with participants and a researcher and were recorded. The questions in the Mass Observation Archive directive were composed by the archive staff and the authors of this book. The responses were individual written replies to the directive from regular archive respondents. All this material was collected in 2006 by the project team.

7 This sort of resistance was also very evident in the focus groups with Catholics run by Grace Reid. These participants often made comments about the way science focused on 'rich, first world, minority problems' and indulgent cosmetic 'cures'. They criticised the exploitation of the image of the desperate sick and replaced it

with an image of greedy, child-like Westerners crying out, 'Oooh, ooh, pick me, I haven't had a child, let's do that' or 'Ooh, ooh, pick me, I've had some booboo that needs to be fixed' (Focus Group 2, Reid 2007)

8 Pope Benedict XVI reaffirmed the position against therapeutic cloning also articulated by the late Pope John Paul II. Pope Benedict XVI's opposition had seen media coverage during the period of the Mass Observation Directive and this was referred to in the directive, both positively and in relation to a view that the Pope (as representing religion) should not take a stance on an area of science.

9 Three out of 175 Mass Observation participants.

10 In contra-distinction to the sterotypes of science-fiction fans as younger males, a large proportion of this mixed gender group had graduate-level education and had studied science at college.

11 Religion can also resource this view – in Grace Reid's Catholic groups, life and death, and indeed suffering, are the gift of God. Seeking to cure Ronald Reagan of his illness was seen as, in their words, 'icky', because Ronald Reagan had led a full and natural life (Focus Group 1, Reid 2007).

12 These developments did not occur in isolation. The public understanding of science has been on the policy agenda in the UK since it appeared in the Royal Society's Bodmer report of 1985.

13 *If* has been nominated for a 'Window on Science' award by the UK/European Public Awareness of Science and Technology (PAWS) initiative 2006. It is also worth noting that *If* was made for BBC 2, which is a television channel with an educational and minority remit in the UK.

14 This occurred in recent coverage of the hybrid-embryo debate in the UK.

15 Our project is part of this proliferation and was funded through the ESRC Genomics Network.

16 The Sci-Art programme has been running successfully since 1996 and it funds collaborative projects for artists and scientists and provides opportunities for dissemination. Over a decade it has financed 124 projects to a total value of £3 million.

8 Conclusion

1 For example, both the European Association for Studies of Science and Technology and the Society for Social Studies of Science held sessions on the 'Hwang affair' at their bi-annual and annual conferences in 2006.

References

Adam, D. (2006) 'Maverick media reveals details of baby cloning experiment'. *Guardian*, 20 July: 17.

Adams, D. (2002) 'Cult from outer space claims to have cloned baby'. *The Times*, 28 December: 1.

Anderson, B. (1983) *Imagined Communities: Reflections on the Origin and Spread of Nationalism*. London and New York: Verso.

Andrews, L. and D. Nelkin (2001) *Body Bazaar: The Market for Human Tissue in the Biotechnology Age*. New York: Crown.

Anker, S. and D. Nelkin (2003) *The Molecular Gaze: Art in the Genetic Age*. Cold Spring Harbor, NY: Cold Spring Harbor Laboratory Press.

Armitt, L. (1991) *Where No Man Has Gone Before: Women and Science Fiction*. London: Routledge.

Balsamo, A. (1999) 'Notes towards a reproductive theory of technology'. In E. A. Kaplan and S. M. Squier (eds) *Playing Dolly: Technocultural Formations, Fantasies, and Fictions of Assisted Reproduction*. New Brunswick, NJ and London: Rutgers University Press.

Balsamo, A. (1996) *Technologies of the Gendered Body: Reading Cyborg Women*. Durham and London: Duke University Press.

Barr, M. S. (1993) *Lost in Space*. Chapel Hill and London: The University of North Carolina Press.

Barron, P. (2005) 'The Newsnight Mission'. http://news.bbc.co.uk/1/hi/programming/newsnight/newsnight25/4198849.stm (23 January).

Barthe, Y. and D. Lindhardt (2006) 'Democratising science and technology: How to be critical?' *Silence, Suffering and Survival: Society for the Social Studies of Science Annual Meeting*. Vancouver.

Bates, B. (2005) 'Public culture and public understanding of genetics: a focus group study'. *Public Understanding of Science* 14(1): 47–65.

Battaglia, D. (2001) 'Multiplicities: An anthroplogist's thoughts on replicants and clones in popular film'. *Critical Inquiry* 27(3): 493–514.

Beck, U. (1992) *Risk Society: Towards a New Modernity*. London: Sage.

Bhabha, H. (1990) *Nation and Narration*. London and New York: Routledge.

Bharadwaj, A. (2007) 'Biosociality and bio-crossings: encounters with assisted conception and embryonic stem cells in India'. In S. Gibbon and C. Novas (eds)

Genetics, Biosociality and the Social Sciences: Making Biologies and Identities. London and New York: Routledge.

Blair, T. (2006) 'Our Nation's future: lecture 4'. London: HMSO. (Also available www.number-10.gov.uk/output/Page10318.asp)

Bloor, D. (1976) *Knowledge and Social Imagery*. London: Routledge and Kegan Paul.

Borger, J. (2002) 'Cult scientists claim first human cloning: Calls for worldwide ban as anger at "mavericks" grows'. *Guardian*, 28 December: 1.

Bowring, F. (2004) 'Therapeutic and reproductive cloning: a critique'. *Social Science and Medicine* 58: 401–409.

Briggs, R. W. and T. J. King (1956) 'Serial Transplantation of Embryonic Nuclei'. *Cold Spring Harbor Symposia on Quantitative Biology* 21: 271–290.

Bromhall, J. D. (1975) 'Nuclear transplantation in the rabbit egg'. *Nature* 258: 719–722

Butler, J. (1990) *Gender Trouble: Feminism and the Subversion of Identity*. New York: Routledge.

Butler, J. (1997) *Excitable Speech: A Politics of the Performative*. London: Routledge.

Butler, J. (2004) *Precarious Life: The Powers of Mourning and Violence*. London: Verso.

Bygrave, M. (2002) 'False dawns in the brave world of new genetics'. *Observer*, 22 December.

Cartwright, L. (1995) *Screening the Body: Tracing Medicine's Visual Culture*. Minnesota: Minnesota University Press.

Center for Genetics and Society (2003) 'Human Cloning, the United Nations and Beyond'. *Newsletter of the Center for Genetics and Society*, January 13. (www.genetics-and-society.org/newsletter/archive/36.html)

Chekar, C. and J. Kitzinger (2007) 'Science, patriotism and constructions of nation and culture'. *New Genetics and Society* 26(3): 249-264.

Chong, S. and D. Normile (2006) 'How young Korean researchers helped unearth a scandal...' *Science* 311(22): 22–25.

Clarke, A. (1998) *Disciplining Reproduction*. Berkeley, LA, and London: University of California Press.

Connor, S. (2002) 'A baby called Eve and a cult that believes in aliens'. *Independent*, 28 December: 3.

Connor, S. (2005a) 'Scientists given right to create baby with two genetic mothers'. *Independent*, 9 September: 18.

Connor, S. (2005b) 'Test-tube baby will have father and two mothers.' *Independent*, 9 September: 23.

Connor, S. and C. Arthur (2004) 'Embryo science: cloning breakthrough opens the door to new treatments – and to a fierce ethical debate.' *Independent*, 13 February: 4–5.

Cookson, C. (2004) 'Leading scientists condemn "cowboy cloners"'. *Financial Times*, 21 January: 5.

Cookson, C. (2005) 'An Asian triumph for therapeutic cloning'. *Financial Times*, 20 May: 14.

Crowe, C. (1990) 'Whose mind over whose matter? Women, in vitro fertilisation and the development of scientific knowledge'. In M. McNeil, I. Varcoe and S. Yearly (eds) *The New Reproductive Technologies*. London: Macmillan.

Cyranoski, D. (2004) 'Korea's stem-cell stars dogged by suspicion of ethical breach'. *Nature* 429: 3.

Daly, M. (1978) *Gyn/Ecology: The Metaethics of Radical Feminism*. London: Beacon Press.

Doane, M. A. (1990) 'Technophilia: technology, representation and the feminine'. In M. Jacobus, E. Fox Keller and S. Shuttleworth (eds) *Body/Politics: Women and the Discourses of Science*. New York: Routledge.

Donawerth, J. (1997) *Frankenstein's Daughters: Women Writing Science Fiction*. Syracuse: Syracuse University Press.

Douglas, M. (1966) *Purity and Danger: An Analysis of Concepts of Pollution and Taboo*. London: Routledge and Kegan Paul.

Downes, M. (2004) '*If* . . . editor's message: you decide the future'. http://news.bbc.co.uk/1/hi/programmes/if/4076517.stm (9 December).

Durant, J., A. Hansen and M. Bauer (1996) 'Public understanding of the new genetics'. In T. Marteau and M. Richards (eds) *The Troubled Helix: Social and Psychological Implications of the New Human Genetics*. Cambridge: Cambridge University Press.

Duster, T. (2003) *Backdoor to Eugenics*. New York and London: Routledge.

Easlea, B. (1987) *Fathering the Unthinkable: Masculinity, Scientists and the Nuclear Arms Race*. London: Pantheon Books.

Editorial (2004a) 'A clone too far'. *The Sunday Times*, 18 January: 20.

Editorial (2004b) 'Science seeks to deliver a baby with three parents'. *Daily Mail*, 18 October: 31.

Fara, P. (2002) *Newton: The Making of Genius*. London: Picador.

Fazackerley, A. (2006) 'Tiny fibs, little fudges and 900 entirely fictional patients'. *The Times Higher Educational Supplement*, 24 February: 19.

Fifty-Ninth General Assemby of the United Nations. (2005) 'General Assembly adopts United Nations Declaration on human cloning by vote of 84–34–37.' www.un.org/News/Press/docs/2005/ga10333.doc.htm (retrieved 24 February 2007).

Firestone, S. (1970) *The Dialectic of Sex*. London: The Women's Press.

Franklin, S. (1990) 'Deconstructing desperateness: the social construction of infertility in popular representations of new reproductive technologies'. In M. McNeil, S. Yearley and I. Varcoe *The New Reproductive Technologies*. London: Macmillan. 200–229.

Franklin, S. (1999) 'What we know and what we don't about cloning and society'. *New Genetics and Society* 18(1): 111.

Franklin, S. (2000) 'Life itself: global nature and the genetic imaginary'. In S. Franklin, C. Lury and J. Stacey *Global Nature, Global Culture*. London: Sage.

Franklin, S. (2003) 'Ethical biocapital: new strategies of cell culture'. In S. Franklin and M. Lock (eds) *Remaking Life and Death: Towards an Anthropology of the Biosciences*. Santa Fe: James Currey Publishers.

Franklin, S. (2005) 'Stem Cells R Us: emergent life forms and the global biological'. In A. Ong and S. Collier (eds) *Global Assemblages: Technology, Politics and Ethics as Anthropological Problems*. New York and London: Blackwell. 59–78.

Franklin, S. (2006) 'Embryonic economies: the double reproductive value of stem cells'. *BioSocieties* 1: 71–90.

Franklin, S., C. Lury and J. Stacey (2000) *Global Nature, Global Culture*. London: Sage.

Fraser, N. (1992) 'Rethinking the public sphere: a contribution to the critique of actually existing democracy'. In C. Calhoun *Habermas and the Public Sphere*. Cambridge, MA: MIT Press. 109–142.

Frayling, C. (2005) *Mad, Bad and Dangerous? The Scientist and the Cinema*. London: Reaktion Books.

Gieryn, T. F. (1999) *Cultural Boundaries of Science: Credibility on the Line*. Chicago: University of Chicago Press.

Gledhill, R. (2004) '"Irresponsible" embryo cloning scientist attacked from all sides'. *The Times*, 19 January: 9.

Goodwin, M. (2006) *Black Markets: The Supply and Demand of Body Parts*. Cambridge: Cambridge University Press.

Gray, A. (2003) *Research Practice for Cultural Studies: Ethnographic Method and Lived Culture*. London: Sage.

Gregory, J. and S. Miller (1998) *Science in Public: Communication, Culture and Credibility*. New York: Basic Books.

Guardian (2006) Saturday Comment and Debate: News Matrix: 07.01.06–13.01.06. *Guardian*, 14 January: 26.

Hall, S. (1997) *Representation: Cultural Representations and Signifying Practices* London: Sage and Open University Press.

Haran, J. (2007) 'Managing the boundaries between maverick cloners and mainstream scientists: the life cycle of a news event in a contested field'. *New Genetics and Society* 26(2): 1–16.

Haraway, D. (1989) *Primate Visions: Gender, Race and Science in the World of Modern Science*. London and New York: Verso.

Haraway, D. (1991) *Simians, Cyborgs and Women: The Reinvention of Nature*. London: Free Association Books.

Haraway, D. (1997) *Modest_Witness@Second_Millenium.FemaleMan©_Meets_Onco Mouse™*. London and New York: Routledge.

Hardt, M. and A. Negri (2001) *Empire*. Cambridge, MA: Harvard University Press.

Hargreaves, I. and G. Ferguson (2000) *Who's Misunderstanding Whom?* Swindon: Economic and Social Research Council.

Hargreaves, I., J. Lewis and T. Speers (2003) *Towards a Better Map: Science, the Public and the Media*. London: ESRC.

Harrell, E. (2006) '"Dolly" scientist cleared of racially harassing and bullying colleague'. *The Scotsman*, 24 August: 13.

Harris, J. (2004) *On Cloning*. London and New York: Routledge.

Hartouni, V. (1997) *Cultural Conceptions: On Reproductive Technologies and the Remaking of Life*. Minneapolis: University of Minnesota Press.

Harvey, O. (2005) 'Regulating stem-cell research and human cloning in an Australian context: an exercise in protecting the status of the human subject'. *New Genetics and Society* 23(2): 125–135.

Haynes, R. D. (1994) *From Faust to Strangelove: Representations of the Scientist in Western Literature*. Baltimore, MD: Johns Hopkins University Press.

Heath, D., R. Rapp and K. S. Taussig (2003) 'Flexible eugenics: discourses of perfectibility and free choice at the end of the 20th century'. In A. Goodman, D. Heath and S. Lindee (eds) *Genetic Nature/Culture: Anthropology and Science Beyond the Two-Culture Divide*. Berkeley: University of California Press.

Helgesen, G. (1995) *Democracy in South Korea: A Political Culture Perspective*. Denmark: NIAS Press.

Henderson, L. and J. Kitzinger (1999) 'The human drama of genetics: "hard" and "soft" media representations of inherited breast cancer'. *Sociology of Health and Illness* 21(5): 560–578.

Henderson, M. (2006a) 'Cloning benefits oversold, says stem-cell scientist'. *The Times*, 18 December: 4.

Henderson, M. (2006b) 'Fraud's the only fool'. *The Times*, 28 January: 5.

Henderson, M. and N. Hawkes (2004) 'Scientists hail human stem-cell breakthrough'. *The Times*, 13 February: 10.

Henig, R. M. (2004) *Pandora's Baby: How the First Test Tube Babies Sparked the Reproductive Revolution*. Cold Spring Harbor: Cold Spring Harbor Laborary Press.

Highfield, R. (2004) 'From science fiction to unproved fact. The race to clone has attracted hoaxers, mavericks and a UFO cult, but even the more reputable attempts have been problematic'. *Daily Telegraph*, 12 August: 4.

Highfield, R. (2004) 'Research will aid the drive for new cures'. *Daily Telegraph*, 12 August: 4.

Highfield, R. (2005a) 'Cloning star who fooled the world. South Korean scientist admits claims were false'. *Daily Telegraph*, 24 December: 4.

Highfield, R. (2005b) 'Designer babies to wipe out diseases approved'. *Daily Telegraph*, 9 September: 1.

Highfield, R. (2005c) 'Disgraced cloning expert blames colleague'. *Daily Telegraph*, 27 December: 2.

Highfield, R. and I. Wilmut (2006) *After Dolly: The Uses and Misuses of Human Cloning*. London: Norton.

Hills, M. (2002) *Fan Cultures*. London and New York: Routledge.

Hills, M. (2004) 'The generic engineering of monstrosity: appropriation of genetics in 1990s "species-level biohorror"'. *MeCCSA*. University of Sussex.

Holliman, R. (2004) 'Media coverage of cloning: a study of media content, production and reception.' *Public Understanding of Science* 13: 107–130.

Horton, R. (2006) 'Response: The cloning fraud case is a scientific success story: A staggering fabrication has been uncovered. This is to our credit, not our shame, says Richard Horton'. *Guardian*, 13 January: 39.

House of Lords (2000) *3rd Report of Science and Technology Select Committee: Science and Society*.

Hughes, E., J. Kitzinger and G. Murdock (2006) 'Risk and the media'. In P. Taylor-Gooby and J. Zinn (eds) *Risk in Social Science*. Oxford: Oxford University Press. 250–270.

Human Fertilisation and Embryology Authority (2006) *Donating Eggs for Research: Safeguarding Donors*. London: HFEA.

Hwang, W. S., S. I. Roh, B. C. Lee, S. K. Kang, D. K. Kwon, S. Kim et al. (2005)

'Patient-specific embryonic stem cells derived from human SCNT blastocysts'. *Science* **308**(5729): 1777–1783.

Hwang, W. S., Y. J. Ryu, J. H. Park, E. S. Park, E. G. Lee, J. M. Koo et al. (2004) 'Evidence of a pluripotent human embryonic stem cell line derived from a cloned blastocyst'. *Science* **303**(5664): 1669–1674.

Iglauer, P. D. (2005) 'Hwang's 40-foot gorilla.' *Korea Times*, 19 December.

Illmensee, K. and P. C. Hoppe (1981)'Nuclear transplantation in mus musculus: developmental potential of nuclei from preimplantation embryos', *Cell* **23**(1):9–18.

Irwin, A. and B. Wynne (1996) *Misunderstanding Science? The Public Reconstruction of Science and Technology*. Cambridge: Cambridge University Press.

Jasanoff, S. (2005) 'In the democracies of DNA: ontological uncertainty and political order in three states'. *New Genetics and Society* 24(2): 139–155.

Jenkins, H. (1992) *Textual Poachers: Television Fans and Participatory Culture*. New York: Routledge.

Jeong, Y. (2006) 'Representation of Korean women's bodies in biomedical technologies: from birth control to stem cell research'. Presentation at the *4S Conference*, Vancouver.

Jha, A. (2004) 'Unexpected results: scientists often think the public will believe anything when it comes to new research, but studies show that this viewpoint is unproven'. *Guardian*. 22 January: 25.

Johnson, R., D. Chambers, E. Tincknell and P. Raghuram (2004) *The Practice of Cultural Studies*. London: Sage.

Jones, H. (2004) 'The cloning process'. *News of the World*, 18 January.

Kay, L. (2000) *Who Wrote the Book of Life? A History of the Genetic Code*. California, CA: Stanford University Press.

Keller, E. F. (1995) *Re-figuring Life: Metaphors of Twentieth Century Biology*. New York: Columbia University Press.

Keller, E. F. (2000) *The Century of the Gene*. Cambridge, MA: Harvard University Press.

Kember, S. (2003) *Cyberfeminism and Artificial Life*. London and New York: Routledge.

Kerr, A. and S. Cunningham-Burley (2000) 'On ambivalence and risk: reflexive modernity and the new human genetics'. *Sociology* 34(2): 283–304.

Kim, E. S. (2006) *Impure Bioethics: Social and Policy Studies of Bioethics Regarding Stem Cell Research in the United States and South Korea*. PhD thesis. Rensselaer Polytechnic Institute, Troy, New York.

Kirby, D. (2003) 'Science consultants, fictional films, and scientific practice'. *Social Studies of Science* 33(2): 231–268.

Kitzinger, J. (2007) 'Framing and frame analysis'. In E. Devereux (ed.) *Media Studies: Key Issues and Debates*. London: Sage. 144–161.

Kitzinger, J. and C. Williams (2005) 'Forecasting science futures: legitimising hope and calming fears in the embryo stem cell debate'. *Social Science & Medicine* 61(3): 731–740.

Kitzinger, J., C. Williams and L. Henderson (2007) 'Science, media and society: the framing of bioethical debates around embryonic stem cell research between 2000 and 2005'. In P. Glasner, P. Atkinson and H. Greenslade (eds) *New Genetics, New Social Formations*. London and New York: Routledge. 204–230.

Klotzko, A. J. (2001) *The Cloning Sourcebook*. Oxford: Oxford University Press.

Klotzko, A. J. (ed.) (2003) 'Voices from Roslin: the creators of Dolly discuss cloning science, ethics, and social responsibility'. *The Cloning Sourcebook*. Oxford: Oxford University Press. 3–27.

Klotzko, A. J. (2004) *A Clone of Your Own?* Oxford: Oxford University Press.

Kolata, G. (1997) 'Scientist reports first cloning ever of adult mammal'. *The New York Times*, 23 February.

Kolata, G. (1998) *Clone: The Road to Dolly and the Path Ahead*. New York: William Morrow.

Kuhn, A. (1990) *Alien Zone: Cultural Theory and Contemporary Science Fiction Cinema*. London: Verso.

Laclau, E. and C. Mouffe (1998) *The Radical Democratic Imaginary*. London and New York: Routledge.

Latour, B. (1986) *Laboratory Life: The Social Construction of Scientific Facts*. New York: Princetown University Press.

Laurance, J. (2004a) 'US doctor searches for British woman to carry cloned baby'. *Independent*, 16 January.

Laurance, J. (2004b) 'Woman is carrying first cloned embryo, claims doctor'. *Independent on Sunday*, 18 January: front page.

Lewenstein, B. (1995) 'From fax to facts: communication in the cold fusion saga'. *Social Studies of Science* 25: 403–436.

Locke, S. (1999) 'Golem science and the public understanding of science: from deficit to dilemma'. *Public Understanding of Science* 8: 75–92.

Lury, C. (2000) 'The united colors of diversity: essential and inessential culture'. In S. Franklin, C. Lury and J. Stacey (eds) *Global Nature, Global Culture*. London: Sage. 146-187.

McNeil, M. (1988) 'Newton as a national hero'. In J. Fauvel, R. Flood, M. Shortland and R. Wilson (eds) *Let Newton Be! A New Perspective on His Life and Works*. Oxford: Oxford University Press.

McNeil, M. (1993) 'New reproductive technologies: dreams and broken promises'. *Science as Culture* 17(3): 483–506.

McRobbie, A. (1998) *British Fashion Design: Rag Trade or Image Industry?* London and New York: Routledge.

Miller, D., J. Kitzinger and P. Beharrell (1998) *The Circuit of Mass Communication: Media Strategies, Representation and Audience Reception in the AIDS Crisis*. London: Sage.

Mitchell, R. and C. Waldby (2006) *Tissue Economies: Blood Organs and Cell Lines in Late Capitalism*. Durham: Duke University Press.

Morton, E. (2004) 'Scientists' cloning success "will cure incurable"'. *Sun*, 13 February: 8–9.

Mulkay, M. (1994) 'The triumph of the pre-embryo: interpretations of the human embryo in parliamentary debate over embyro research'. *Social Studies of Science* 24(4): 611–639.

Mulkay, M. (1997) *The Embryo Research Debate: Science and the Politics of Reproduction*. Cambridge and New York: Cambridge University Press.

Nahman, M. (2006) 'A different mix: hybridity in Israeli ova donation'. *Science as Culture* 15(3): 199–213.

Nature (2006) 'Editorial'. *Nature* 439(7023): 117-18. (12 January)
Nelkin, D. (1987) *Selling Science: How the Press Covers Science and Technology*. New York: W. H. Freeman and Co.
Nelkin, D. and M. S. Lindee (1995) *The DNA Mystique: The Gene as Cultural Icon*. New York: Freeman.
Nerlich, B. and D. D. Clarke (2003) 'Anatomy of a media event: how arguments clashed in the 2001 human cloning debate'. *New Genetics and Society* 22(1): 43–59.
Nerlich, B., D. D. Clarke and R. Dingwall (2001) 'Fictions, fantasies, and fears: the literary foundations of the cloning debate'. *Journal of Literary Semantics* 30: 37–52.
Nerlich, B., R. Dingwall and P. Martin (2004) 'Genetic and genomic discourses at the dawn of the 21st century'. *Discourse & Society* 15(4): 363–368.
Nerlich, B., R. Dingwall and D. D. Clarke (2002) 'The book of life: how the completion of the Human Genome Project was revealed to the public'. *Health: An Interdisciplinary Journal for the Social Study of Health, Illness and Medicine* 6(4): 445–469.
Nerlich, B. and I. Hellsten (2004) 'Genomics: shifting metaphorical landscapes between 2000 and 2003'. *New Genetics and Society* 23(3): 255–268.
Noble, D. (1992) *A World Without Women: The Christian Clerical Culture of Western Science*. London: Oxford University Press.
Normile, D., G. Vogel and J. Couzin (2006) 'South Korean team's remaining human stem cell claim demolished'. *Science* 311(5758): 156–157.
Norsigian, J. (2005) 'Egg donation for IVF and stem cell research: time to weigh the risks to women's health'. *Differentakes* 33 (spring): 1–4. (Also available at http://popdev.hampshire.edu/projects/dt/dt33.php)
Nussbaum, M. C. and C. R. Sunstein (1999) *Clones and Clones: Facts and Fantasies about Human Cloning*. New York and London: W. W. Norton & Company.
Paget, D. (1998) *No Other Way to Tell It: Dramadoc/docudrama on Television*. Manchester: Manchester University Press.
Paik, Y. G. (2006) 'Beyond bioethics: the globalized reality of ova trafficking and the possibility of feminist intervention'. Presentation at the International Forum on the Human Rights of Women and Biotechnology, 21 September 2006. Seoul Women's Plaza, Seoul, Korea.
Parrinder, P. (1980) *Science Fiction: Its Criticism and Teaching*. London and New York: Methuen.
Parrinder, P. (1997) 'Eugenics and utopia: sexual selection from Galton to Morris.' *Utopian Studies* 8(2): 1–12.
Parry, S. (2006) '(Re)constructing embryos in stem cell research: exploring the meaning of embryos for people involved in fertility treatments'. *Social Science & Medicine* 62(10): 2349–2359.
Parry, S. (2003) 'The politics of cloning: mapping the rhetorical convergence of embryos and stem cells in parliamentary debates'. *New Genetics and Society* 22(2): 177–200.
Petchesky, R. (1987) 'Foetal images: the power of visual culture in the politics of reproduction'. In M. Stanworth (ed.) *Reproductive Technologies: Gender, Motherhood and Medicine*. Minneapolis: University of Minnesota Press.
Petersen, A. (2002) 'The new genetics and the media'. In A. Petersen and R. Bunton (eds) *The New Genetics and the Public's Health*. London and New York: Routledge.

Pfeffer, N. (1993) *The Stork and the Syringe: A Political History of Reproductive Medicine*. London: Polity.

Pfeffer, N. and A. Woolett (1983) *The Experience of Infertility*. London: Virago Press.

Poon, P. N. (2000) 'Evolution of the clonal man: inventing science unfiction'. *Journal Of Medical Humanities* 21(3): 159–173.

Probyn, E. (2005) *Blush: Faces of Shame*. Minneapolis: Minnesota University Press.

Rabinow, P. and N. Rose (2003) 'Thoughts on the concept of biopower today'. www.molsci.org/files/Rose_Rabinow_Biopower_Today.pdf.

Rael, C. (2001) *Yes to Human Cloning: Eternal Life Thanks to Science*. Trans. M. Wenner. Raelian Foundation.

Randerson, J. (2005) 'Rise and fall of clone king who doctored stem-cell research: Scientists' fury at setback for research after fraud: Fairytale ascent from rags to riches ends in disgrace'. *Guardian*, 24 December: 6.

Reid, G. (2007) 'Replicating opinions? Audience responses to a television dramadoc about human cloning.' Draft Ph.D. thesis, Cardiff University.

Reid, G. T. (2004) 'Clonesequences: social representations of cloning risks and benefits'. *Faculty of Communication and Culture*. Calgary, Alberta: University of Calgary.

Revill, J. (2004) 'Scientists pour scorn on doctor's human clone boast: uproar at news conference as US fertility maverick drops genetic bombshell'. *Observer*, 18 January: 5.

Roberts, C. and K. Throsby (2006) 'Paid to share: IVF patients, eggs and stem cell research'. Paper presented at *Feminist Issues in Contemporary UK Biopolitics – Sourcing eggs for biomedical research*. CESAGen, Cardiff University, 20 November 2006.

Roof, J. (1996) *Reproductions of Reproduction: Imaging Symbolic Change*. London and New York: Routledge.

Rose, N. (2001) 'The politics of life itself'. *Theory, Culture and Society* 18(6): 1–30.

Rose, S. (2006) 'Defending Dolly.' *The Lancet* 368(9544): 1319.

Sang-Hun, C. (2006) 'Lesson in South Korea: stem cells aren't cars or chips'. *The New York Times*, 11 January: 16.

Sargent, P. (1976) *Cloned Lives*. Greenwich, CN: Fawcett Gold Medal.

Schwarz, C. (1993) *The Chambers Dictionary*. Edinburgh: Chambers.

Science Media Centre (2002) *Science Media Centre: Consultation Report*. Science Media Centre.

Science Media Centre (2003) *Peer review in a nutshell*. Science Media Centre.

Science Media Centre (2004) *Open Letter to News Editors*. Science Media Centre.

Sexton, S. (2000) 'How to talk about cloning without talking about cloning: public discourse in the UK'. *Biomedical Ethics* 5(3).

Sexton, S. (1999) 'If cloning is the answer, what was the question? Power and decision-making in the geneticisation of health' *Corner House Briefing 16* (www.thecornerhouse.org.uk/item.shtml?x=51972, accessed 24 June 2007)

Shannon, C. and W. Weaver (1963) *The Mathematical Theory of Communication*. Urbana: University of Ilinois Press.

Shapin, S. and S. Schaffer (1989 (1985) *Leviathan and the Air-pump: Hobbes, Boyle, and the Experimental Life*. Princeton, N J: Princeton University Press.

Shildrick, M. (1997) *Leaky Bodies and Boundaries: Feminism, Postmodernism and Bioethics*. New York: Routledge.

Showalter, E. (1985) *The Female Malady: Women, Madness, and English Culture, 1830–1980*. New York: Pantheon Books.

Silver, L. (2001) 'What are clones? They're not what you think they are'. *Nature* 412(5 July): 21.

Simon, B. (2001) 'Public science: media configuration and closure in the cold fusion controversy'. *Public Understanding of Science* 10(4): 383–402.

Simon, B. (2002) *Undead Science*. New Brunswick, NJ and London: Rutgers University Press.

Sobchack, V. (1997) *Screening Space: The American Science Fiction Film*. New Brunswick, NJ and London: Rutgers University Press.

Song, J.-A. (2005) 'Dismay in South Korea at national scientific hero's fake stem cell research'. *The Financial Times*, 27 December.

Spivak, C. (1988) 'Can the subaltern speak?' In C. Nelson and L. Grossberg (eds) *Marxism and the Interpretation of Culture*. Chicago: University of Ilinois Press. 271–313.

Springer, C. (1996) *Electronic Eros: Bodies and Desire in the Post Industrial Age*. Austin: University of Texas Press.

Stabile, C. (1994) *Feminism and the Technological Fix*. Manchester: Manchester University Press.

Stacey, J. (2003) 'She is not herself: the deviant relations of *Alien Resurrection*'. *Screen* 44(3): 251–276.

Stacey, J. (2005) 'Masculinity, masquerade, and genetic impersonation: Gattaca's queer visions'. *Signs: Journal of Women in Culture and Society* 30: 1851–1877.

Stacey, J. (forthcoming) *The Cinematic Life of the Gene*. Raleigh: Duke University Press.

Stern, M. (2004) '*Jurassic Park* and the moveable feast of science'. *Science as Culture* 13(3): 347–372.

Suvin, D. (1979) *Metamorphoses of Science Fiction: On the Poetics and History of a Literary Genre*. New Haven and London: Yale University Press.

Tae-gyu, K. (2003) 'Professor confident of commercializing mad cow disease-free clones.' *The Korea Times*, 11 December.

Tae-gyu, K. (2005) 'Long and bumpy road to doctorate: Prof. Hwang: Move heart of heaven (1)'. *The Korea Times*, 28 May.

Taussig, K.-S., R. Rapp and D. Heath (2003) 'Flexible eugenics: technologies of the self in the age of genetics'. In A. Goodman, D. Heath and S. Lindee (eds) *Genetic Nature/Culture: Anthropology and Science beyond the Two-Culture Divide*. Berkeley, CA: University of California Press.

Templeton, S.-K. (2006) 'Cloning expert quits country in row with partner'. *The Sunday Times*, 15 January: 4.

Thacker, E. (2002) 'Biohorror/Biotech'. *Paradoxa* 17:109-129.

Thacker, E. (2004) *Biomedia*. Minnesota: University of Minnesota Press.

Thacker, E. (2005) *The Global Genome: Biotechnology, Politics, and Culture*. Cambridge, MA: MIT Press.

Thompson, C. (2005) *Making Parents: The Ontological Choreography of Reproductive Technologies*. Cambridge, MA: MIT Press.

Throsby, K. (2004) *When IVF Fails: Feminism, Infertility and the Negotiation of Normality*. London: Palgrave Macmillan.

Thurlbeck, N. (2004) 'Dr Frankenstein'. *News of the World*. London: 15.

Turney, J. (1998) *Frankenstein's Footsteps: Science, Genetics and Popular Culture*, New Haven, CT: Yale University Press.

Usdin, S. (2004) 'Prop. 71: promises to keep'. *BioCentury, The Bernstein Report on BioBusiness* (www.genetics-and-society.org/resources/items/20041108_biocentury_usdin.html)

Utton, T. (2004) 'Playing God? Ethical storm as world's first human clone is created'. *Daily Mail*, 13 February: 1, 5.

van Dijck, J. (1998) *Imagenation: Popular Images of Genetics*. London: Macmillan.

Waldby, C. (2000) *The Visible Human Project: Informatic Bodies and Posthuman Medicine*. London and New York: Routledge.

Waldby, C. (2002) 'Stem cells, tissue cultures and the production of biovalue'. *Health: An Interdisciplinary Journal for the Social Study of Health, Illness and Medicine* 6(3): 305–323.

Waldby, C. and R. Mitchell (2006) *Tissue Economies: Blood, Organs and Cell Lines in Late Capitalism (Science & Cultural Theory)* Durham, MC: Duke University Press.

Watson, J. D. (2000) 'Moving towards the clonal man: is this what we want'. In J. D. Watson (ed.) *A Passion for DNA*. Oxford: Oxford University Press: 83–90.

Webber, H. J. (1903) 'New horticultural and agricultural terms', *Science* **28**: 501–503.

Weingart, P., C. Muhl and P. Pansegrau (2003) 'Of power maniacs and unethical geniuses: science and scientists in fiction film'. *Public Understandings of Science* 12: 279–287.

Wellcome (2006) *Strategic plan: 2005–2010 making a difference*. London: Wellcome Trust. (Also available www.wellcome.ac.uk/strategy/strategicplan/)

Wheldon, J. (2004) 'The clone cure: hope of beating Alzheimer's as 30 human embryos are created'. *Daily Express*, 13 February: 9.

Willadsen, S. M. (1986) 'Nuclear transplantation in sheep embryos'. *Nature* **320**(6057): 63–5.

Williams, C., J. Kitzinger and L. Henderson (2003) 'Envisaging the embryo in stem cell research: rhetorical strategies and media reporting of the ethical debates'. *Sociology of Health and Illness* 25(7): 793–814.

Wilmut, I., K. Campbell and C. Tudge (2000a) *The Second Creation: Dolly and the Age of Biological Control*. New York: Farrar, Strauss and Giroux.

Wilmut, I. and R. Highfield (2006) *After Dolly: The Uses and Misuses of Human Cloning*. London: Little, Brown.

Wilmut, I., A. E. Schnieke, J. McWhir, A. J. Kind and K. H. S. Campbell (2000b) Appendix: The Letter to *Nature* Announcing Dolly's Birth. In Wilmut, K. Campbell and C. Tudge (eds) *The Second Creation: Dolly and the Age of Biological Control*. New York: Farrar, Strauss and Giroux. 307–313.

Winston, R. (2005) 'Presidential address to the BA Festival of Science'. BA Festival of Science, Dublin, 5 September.

Wohn, Y. (2006) 'Seoul National University dismisses Hwang'. *Science* 311: 1695.

Wood, A. (2002) *Technoscience in American Film: Beyond Science Fiction*. Manchester: Manchester University Press.

Woods, R., J. Leake and A. Nathan (2004) 'Driven by his ego: the clone race maverick'. *The Sunday Times*, 18 January: 7.

Wynne, B. (1996) 'Misunderstood misunderstandings: social identities and public uptake of science'. In A. Irwin and B. Wynne (eds) *Misunderstanding Science? The Public Reconstruction of Science and Technology*. Cambridge: Cambridge University Press. 19–46.

Zavos, P. M. and K. Illmensee (2006) 'Possible therapy of male infertility by reproductive cloning: one cloned human 4-cell embryo'. *Archives of Andrology* 52: 243–254.

Index

For Product Safety Concerns and Information please contact our EU representative GPSR@taylorandfrancis.com
Taylor & Francis Verlag GmbH, Kaufingerstraße 24, 80331 München, Germany

www.ingramcontent.com/pod-product-compliance
Ingram Content Group UK Ltd.
Pitfield, Milton Keynes, MK11 3LW, UK
UKHW020010200926
475702UK00014B/1803
* 9 7 8 0 4 1 5 7 5 9 6 6 3 *